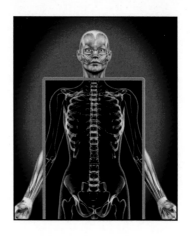

HUMAN BODY - BIG BOOK
Human Body Series

• • • • • • • • • • • • • • • • • • •

Written by Susan Lang

GRADES 5 - 8
Reading Levels 3 - 4

Classroom Complete Press

P.O. Box 19729
San Diego, CA 92159
Tel: 1-800-663-3609 / Fax: 1-800-663-3608
Email: service@classroomcompletepress.com

www.classroomcompletepress.com

ISBN-13: 978-1-55319-381-4
ISBN-10: 1-55319-381-4

© 2007

Critical Thinking Skills

Human Body

Cells, Skeletal System & Muscular System
Senses, Nervous System & Respiratory System
Circulatory, Digestive, Excretory & Reproductive Systems
Human Body - Big Book

Skills For Critical Thinking	Reading Comprehension								Hands-on Activities
	Section 1	Section 2	Section 3	Section 4	Section 5	Section 6	Section 7	Section 8	
LEVEL 1 Knowledge									
• List Details/Facts	✓	✓	✓	✓	✓	✓	✓	✓	✓
• Recall Information	✓	✓	✓	✓	✓	✓	✓	✓	✓
• Match Vocabulary to Definitions	✓	✓	✓	✓	✓	✓	✓	✓	
• Define Vocabulary				✓	✓	✓		✓	
• Label Diagrams	✓	✓	✓	✓	✓			✓	✓
• Recognize Validity (T/F)			✓			✓	✓	✓	
LEVEL 2 Comprehension									
• Demonstrate Understanding	✓	✓	✓	✓	✓	✓	✓	✓	✓
• Explain Scientific Causation			✓	✓		✓	✓	✓	
• Describe	✓	✓	✓	✓	✓	✓	✓	✓	✓
• Classify into Scientific Groups	✓	✓	✓	✓				✓	
LEVEL 3 Application									
• Application to Own Life	✓	✓	✓	✓	✓	✓	✓		✓
• Organize and Classify Facts	✓		✓	✓					
LEVEL 4 Analysis									
• Make Inferences		✓	✓	✓	✓	✓	✓	✓	✓
• Draw Conclusions Based on Facts Provided		✓	✓	✓	✓	✓		✓	✓
• Classify Based on Facts Researched	✓	✓	✓	✓		✓		✓	✓
LEVEL 5 Synthesis									
• Compile Research Information	✓	✓	✓	✓	✓	✓	✓	✓	✓
• Design and Application						✓			✓
• Create and Construct									✓
• Ask questions	✓	✓	✓		✓	✓		✓	✓
LEVEL 6 Evaluation									
• State and Defend an Opinion				✓		✓	✓		
• Defend Selections and Reasoning				✓		✓	✓		✓

Based on Bloom's Taxonomy

Contents

• • • • • • • • • • • • • • • •

Contents

✔ **6 BONUS** Activity Pages! **Additional worksheets for your students**

FREE!

• Go to our website: **www.classroomcompletepress.com/bonus**
• Enter item CC4516 – Cells, Skeletal System & Muscular System
• Enter pass code CC4516D for Activity Pages.

✔ **6 BONUS** Activity Pages! **Additional worksheets for your students**

FREE!

• Go to our website: **www.classroomcompletepress.com/bonus**
• Enter item CC4517 – Senses, Nervous System & Respiratory System
• Enter pass code CC4517D for Activity Pages.

✔ **6 BONUS** Activity Pages! **Additional worksheets for your students**

FREE!

• Go to our website: **www.classroomcompletepress.com/bonus**
• Enter item CC4518 – Circulatory, Digestive, Excretory
  & Reproductive Systems
• Enter pass code CC4518D for Activity Pages.

Assessment Rubric

Human Body

Student's Name: _____ Assignment: _____ Level: _____

	Level 1	Level 2	Level 3	Level 4
Understanding Concepts	Demonstrates a limited understanding of concepts. Requires teacher intervention.	Demonstrates a basic understanding of concepts. Requires little teacher intervention.	Demonstrates a good understanding of concepts. Requires no teacher intervention.	Demonstrates a thorough understanding of concepts. Requires no teacher intervention.
Analysis and Application of Key Concepts	Limited application and interpretation in activity responses	Basic application and interpretation in activity responses	Good application and interpretation in activity responses	Strong application and interpretation in activity responses
Creativity and Imagination	Limited creativity and imagination applied in projects and activities	Some creativity and imagination applied in projects and activities	Satisfactory level of creativity and imagination applied in projects and activities	Beyond expected creativity and imagination applied in projects and activities
Application of Own Interests	Limited application of own interests in independent or group environment	Basic application of own interests in independent or group environment	Good application of own interests in independent or group environment	Strong application of own interests in independent or group environment

STRENGTHS:

WEAKNESSES:

NEXT STEPS:

Teacher Guide

Our resource has been created for ease of use by both TEACHERS and STUDENTS alike.

Introduction

This resource provides ready-to-use information and activities for remedial students in grades five to eight. Written to grade using simplified language and vocabulary, **science** concepts are presented in a way that makes them accessible and easier to understand. Comprised of reading passages, student activities and mini posters, our resource can be used effectively for whole-class, small group and independent work.

How Is Our Resource Organized?

STUDENT HANDOUTS

Reading passages and **activities** (*in the form of reproducible worksheets*) make up the majority of our resource. The reading passages present important grade-appropriate information and concepts related to the topic. Included in each passage are one or more embedded questions that ensure students are actually reading and understanding the content.

For each reading passage there are BEFORE YOU READ activities and AFTER YOU READ activities. As with the reading passages, the related activities are written using a remedial level of language.

- The BEFORE YOU READ activities prepare students for reading by setting a purpose for reading. They stimulate background knowledge and experience, and guide students to make connections between what they know and what they will learn. Important concepts and vocabulary are also presented.

- The AFTER YOU READ activities check students' comprehension of the concepts presented in the reading passage and extend their learning. Students are asked to give thoughtful consideration of the

reading passage through creative and evaluative short-answer questions, research, and extension activities.

Hands-on activities are included to further develop students' thinking skills and understanding of the concepts. The **Assessment Rubric** (*page 5*) is a useful tool for evaluating students' responses to many of the activities in our resource. The **Comprehension Quizzes** (*pages 50, 93, 136*) can be used for either a follow-up review or assessment at the completion of the unit.

PICTURE CUES

Our resource contains three main types of pages, each with a different purpose and use. A **Picture Cue** at the top of each page shows, at a glance, what the page is for.

 Teacher Guide
- Information and tools for the teacher

 Student Handout
- Reproducible worksheets and activities

EZ✓ **Easy Marking™ Answer Key**
- Answers for student activities

EASY MARKING™ ANSWER KEY
Marking students' worksheets is fast and easy with this **Answer Key**. Answers are listed in columns – just line up the column with its corresponding worksheet, as shown, and see how every question matches up with its answer!

Every question matches up with its answer!

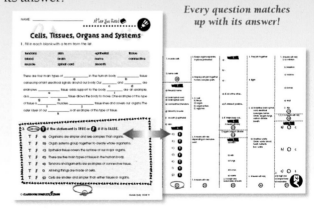

Bloom's Taxonomy

Our resource is an effective tool for any SCIENCE PROGRAM.

Bloom's Taxonomy* for Reading Comprehension

The activities in our resource engage and build the full range of thinking skills that are essential for students' reading comprehension and understanding of important science concepts. Based on the six levels of thinking in Bloom's Taxonomy, and using language at a remedial level, information and questions are given that challenge students to not only recall what they have read, but to move beyond this to understand the text and concepts through higher-order thinking. By using higher-order skills of application, analysis, synthesis and evaluation, students become active readers, drawing more meaning from the text, attaining a greater understanding of concepts, and applying and extending their learning in more sophisticated ways.

Our resource, therefore, is an effective tool for any Science program. Whether it is used in whole or in part, or adapted to meet individual student needs, our resource provides teachers with essential information and questions to ask, inspiring students' interest, creativity, and promoting meaningful learning.

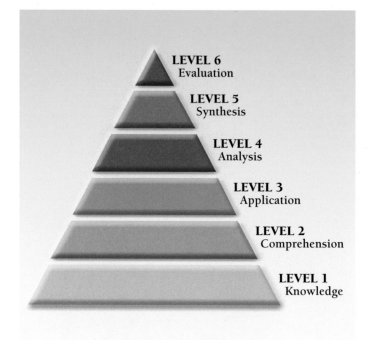

BLOOM'S TAXONOMY: 6 LEVELS OF THINKING

Bloom's Taxonomy is a tool widely used by educators for classifying learning objectives, and is based on the work of Benjamin Bloom.

Vocabulary

Cells, Skeletal System & Muscular System
Senses, Nervous System & Respiratory System
Circulatory, Digestive, Excretory & Reproductive Systems
Human Body - Big Book

abdomen	absorb	acid	alveoli
anvil	aorta	arteries	auricle
balance	bladder	blood vessel	bloodstream
brain	brain stem	branches	bronchial tubes
bundle	bundled	calcium	capillary
carbondioxide	cardiac	cartilage	cell membrane
cerebellum	cerebrum	chamber	chemical

Vocabulary

chemicals	churn	circulate	**circulatory system**
clot	cochlea	coiled	colon
communicate	connected	**connective tissue**	**contracting**
cough	cytoplasm	deoxygenated	diameter
digestion	**digestive system**	ear canal	eardrum
egg cell	**electrical signal**	enzymes	epiglottis
epithelial tissue	esophagus	estrogen	**excretory system**
exhale	expand	fertilization	fetus
filter	germs	hammer	hormone
immunity	indigestible	inhale	**inner ear**
involuntary	iris	iron	**joint rotation**
kidneys	large intestine	liquid	liver
lungs	**lysosomes**	**marrow**	**middle ear**
mineral	**mitochondria**	moist	moisten
motor nerves	mucus	**multicellular**	**muscle fiber**
muscle tissue	**muscular system**	nasal cavity	navel
nerve tissue	**nervous system**	neurons	**nucleus**
nutrients	optic nerve	outer ear	**oxygen**
oxygenated	pancreas	**particle**	perspiration
plasma	platelet	poison	pores
posture	**pressure**	produces	pupil
receptor	red blood cells	**reproductive system**	retina
saliva	secrete	sensations	sensory nerves
small intestine	solid waste	**specialized cells**	species
sperm cell	**spinal cord**	starchy	stirrup
stomach	**striated**	surface	survival
testosterone	**tongue**	toxic	trachea
transfer	umbilical cord	**unicellular**	urine
veins	vertebrae	vibration	**voluntary**
white blood cells	windpipe		

NAME: _____

Cells – The Building Blocks of Life

1. **Complete each sentence with a word from the list. Use a dictionary to help you.**

unicellular organisms   multicellular organisms   specialized
bacteria   cells   microscope

a) Every living thing is made up of _____. That is why they are called the building blocks of life.

b) Some living things are very simple. The ones that are only one cell in size are called

_____.

c) _____ are an example of unicellular organisms.

d) Humans and frogs are an example of _____.

e) Most cells are very small. We have to use a _____ to be able to see them.

2. **Use the cell shapes below to list anything you already know about cells and some questions you have about cells.**

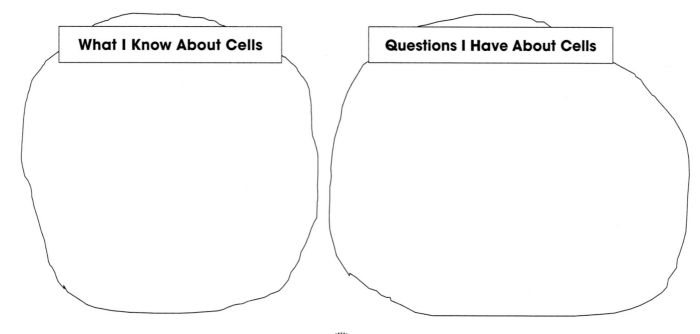

What I Know About Cells

Questions I Have About Cells

NAME: _____

Cells – The Building Blocks of Life

 Cells are called the building blocks of life because every living thing in the world is made of cells.

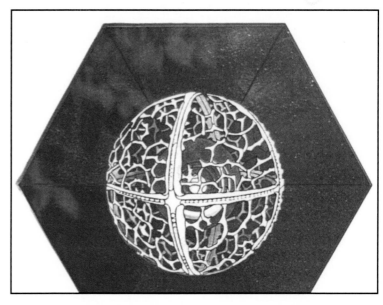

Unicellular Organisms

Some living things are very simple and are only one cell in size. These are called **unicellular organisms.** This one cell is able to do all the things needed to keep the organism healthy and alive. The cell can move, eat, breathe, remove waste and reproduce. There are many unicellular organisms in the world but most are far too small to see without a **microscope. Amoeba** and **bacteria** are examples of unicellular organisms. The largest unicellular organism is the ostrich egg!

Multicellular Organisms

Almost all of the living things we see around us are **multicellular organisms.** They are made of more than one cell. All plants and animals, including humans, are multicellular organisms. Some multicellular organisms are made up of only a few cells but most are made of **billions** of cells. A human baby is born with 26 billion cells, but by the time it is an adult it will be made of close to 100 trillion cells!

Cells that make up multicellular organisms are **specialized.** They do only certain jobs and need all the other cells to do their own specialized jobs too. Working together, all the cells keep the organism alive and healthy. For example, the cells in our eyes help us see but cannot help us breathe. We need the cells in our lungs for that.

STOP

Look around you. List FIVE things you see that are made of cells. Tell whether each thing is UNICELLULAR or MULTICELLULAR.

_____

_____

_____

NAME: _____

Cells – The Building Blocks of Life

1. Fill in each blank with a term from the list.

alive	amoeba	specialized	multicellular organisms
microscope	billions	unicellular organisms	bacteria
humans	cell	different	

Some living things are very simple and are only one _____ in size. These are

a

called _____. These are very small and in most cases can only be seen with

b

a _____. Two examples of unicellular organisms are _____ and

c   d

_____. _____ make up most of the living things that we can

e   f

see around us. All plants and animals, including _____, are multicellular

g

organisms. They get their name because they are made of more than one cell. This is

one way that they are _____ from unicellular organisms. Most multicellular

h

organisms are made of _____ of cells. These cells are also all very _____.

i   j

This means they have certain jobs to help keep the organism _____.

k

2. a) <u>Underline</u> the terms and ideas that describe *unicellular* organisms.

simple	human	bacteria	made of billions of cells	ostrich egg
the cells are specialized	tree	made of only one cell	amoeba
so small you need a microscope to see it				

b) Circle the terms and ideas that describe *multicellular* organisms.

simple	human	bacteria	made of billions of cells	ostrich egg
the cells are specialized	tree	made of only one cell	amoeba
so small you need a microscope to see it				

Cells - The Building Blocks of Life

3. **Why are cells called the building blocks of life?**

_____

4. **How can a unicellular organism survive when it is only one cell in size?**

_____

_____

5. **Which statements describe multicellular organisms and which statements describe unicellular organisms? Color in the cells that match the statements.**

Statement	Unicellular Organism	Multicellular Organism
a) These living things are very simple	◯	◯◯
b) Every plant and animal is this	◯	◯◯
c) The cells of this organism are specialized	◯	◯◯
d) This organism is usually too small to see without a microscope	◯	◯◯
e) Amoeba and bacteria are examples	◯	◯◯
f) You are an example	◯	◯◯

Research

6. **Cancer** is a disease that affects humans. It is caused by the uncontrolled growth and spread of cells in the body. Your task is to investigate cancer in humans. Collect information about **cancer cells** (What do they look like? How do they grow? How do they harm the human body?). Then, find out some of the most common **treatments.** Write your findings in a one-page report. Include pictures or illustrations of cancer cells in your report.

7. **Tap water** is treated so that humans will be safe from harmful unicellular organisms that might live in it. Find out where the tap water in your home comes from (i.e., a nearby lake? an underground reservoir?). Then find out how the tap water in your area is treated. List some of the main unicellular organisms that this treatment protects you from. You can look for your information on the Internet. Or, you may need to ask your teacher to help you contact someone who works for your town or city water treatment plant.

Human Fetus

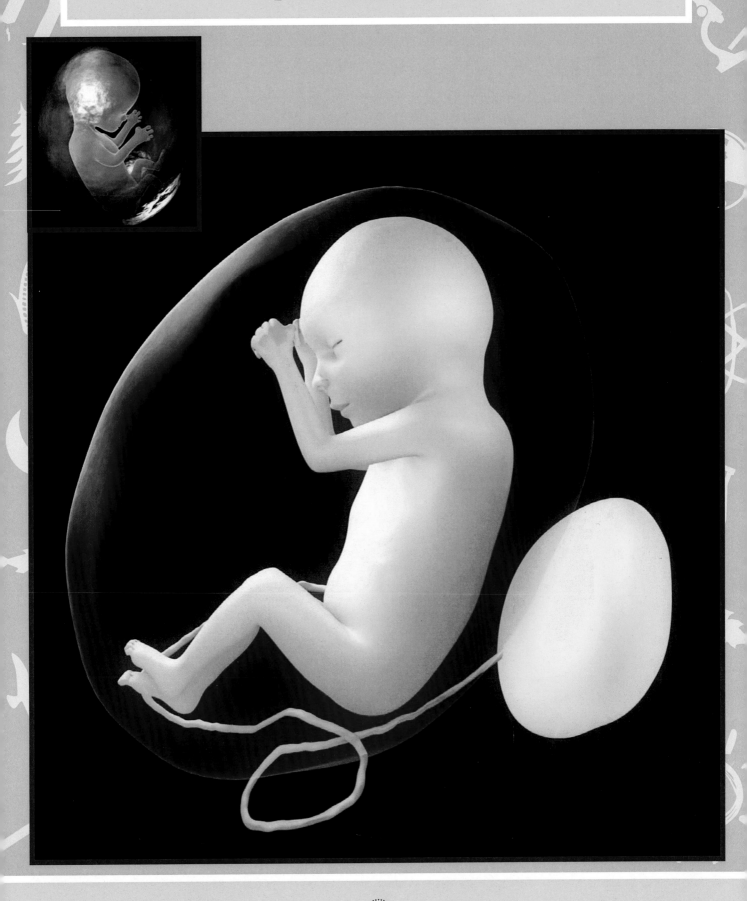

Human Body CC4519

Reproductive System

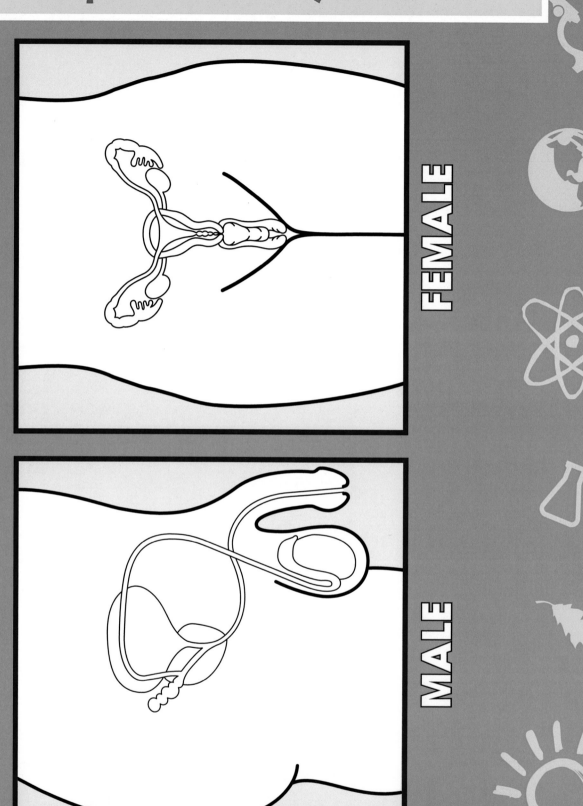

FEMALE

MALE

Respiratory System

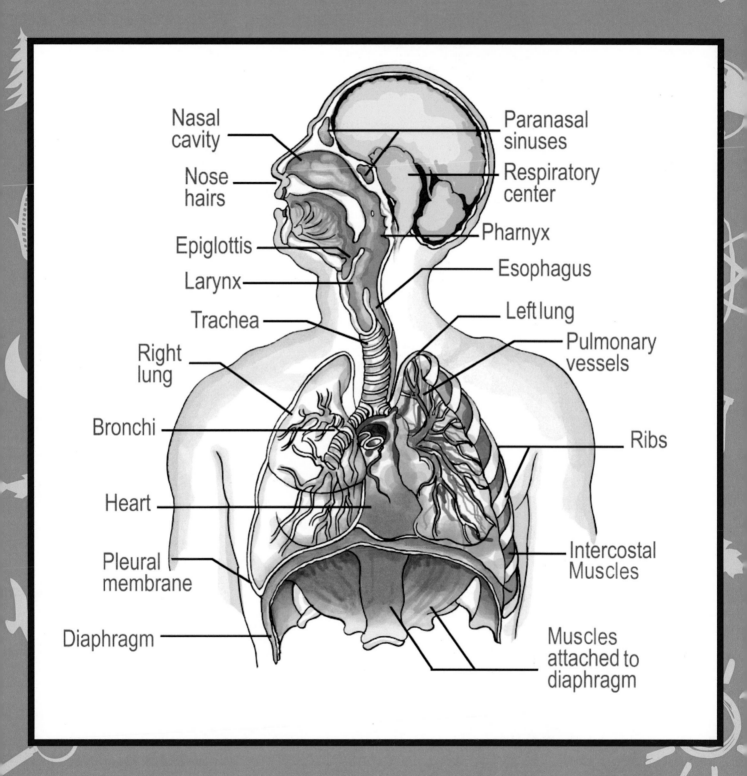

Nasal cavity

Nose hairs

Epiglottis

Larynx

Trachea

Right lung

Bronchi

Heart

Pleural membrane

Diaphragm

Paranasal sinuses

Respiratory center

Pharnyx

Esophagus

Left lung

Pulmonary vessels

Ribs

Intercostal Muscles

Muscles attached to diaphragm

Nervous System

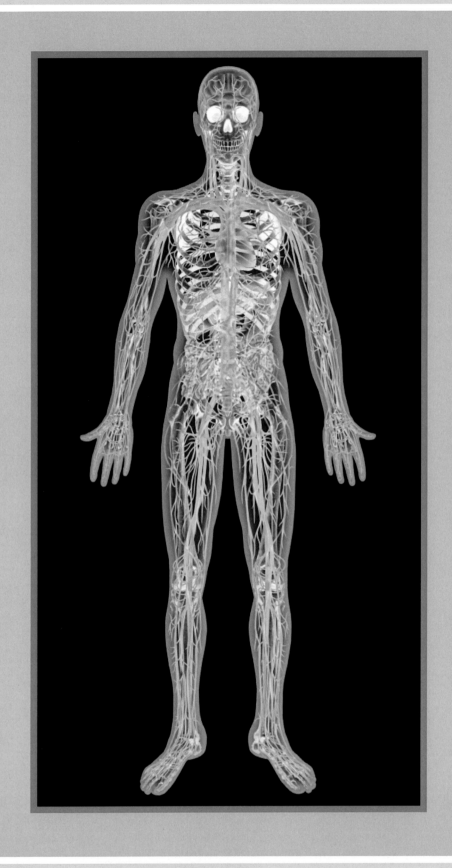

Spinal Cord

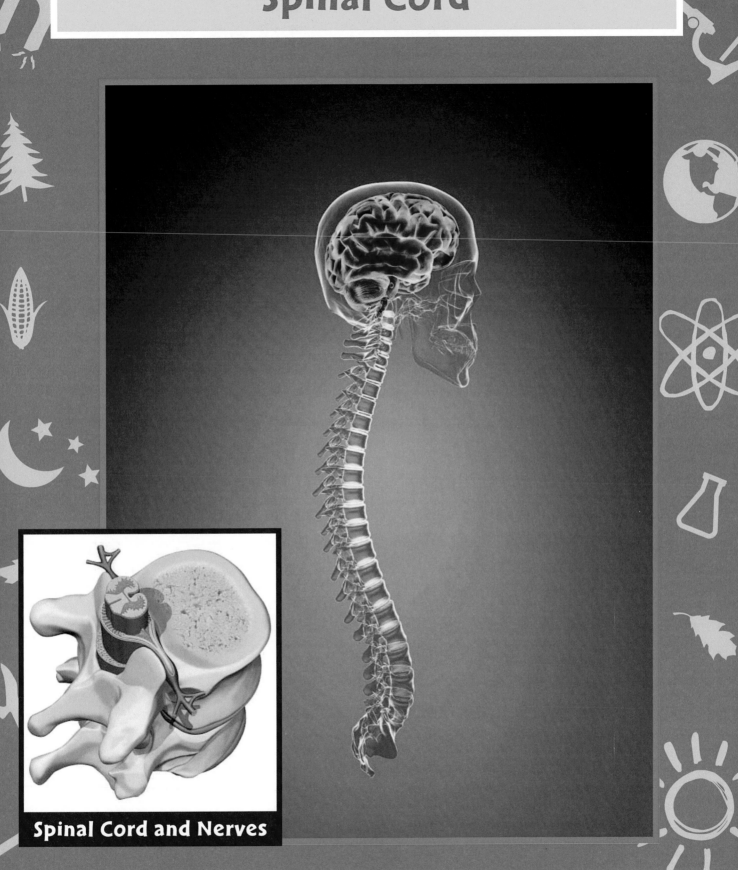

Spinal Cord and Nerves

Parts of the Brain

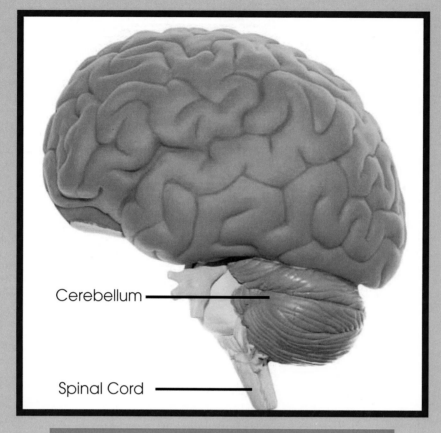

Cerebellum

Spinal Cord

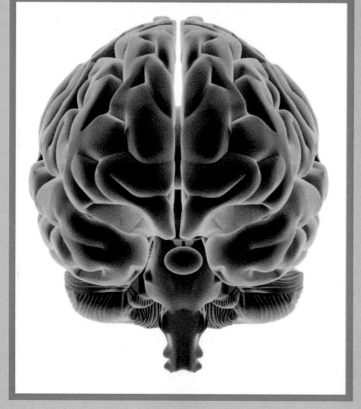

Parts of the Eye

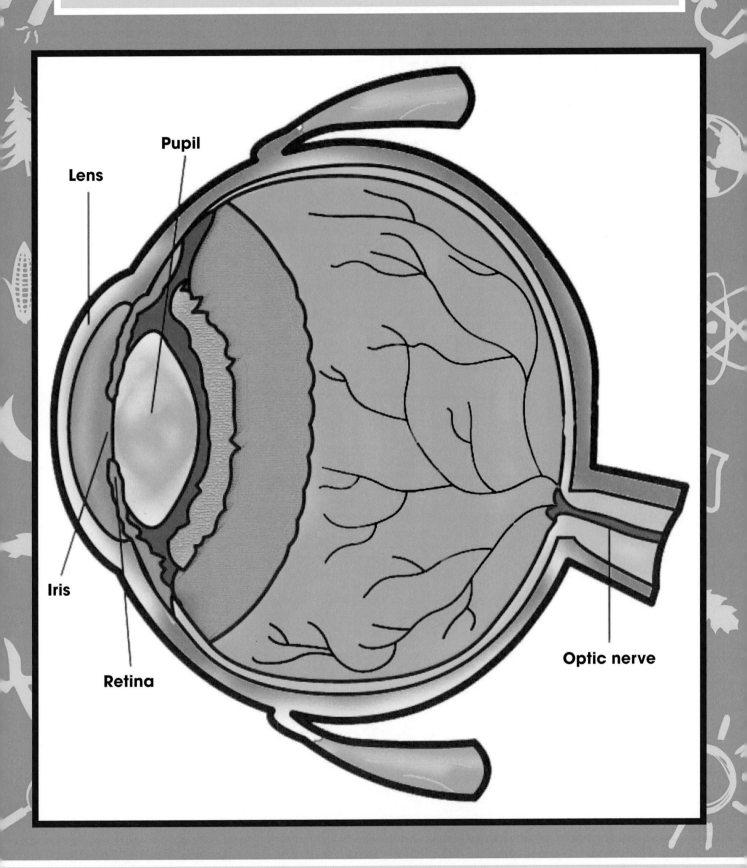

Lens

Pupil

Iris

Retina

Optic nerve

Parts of the Ear

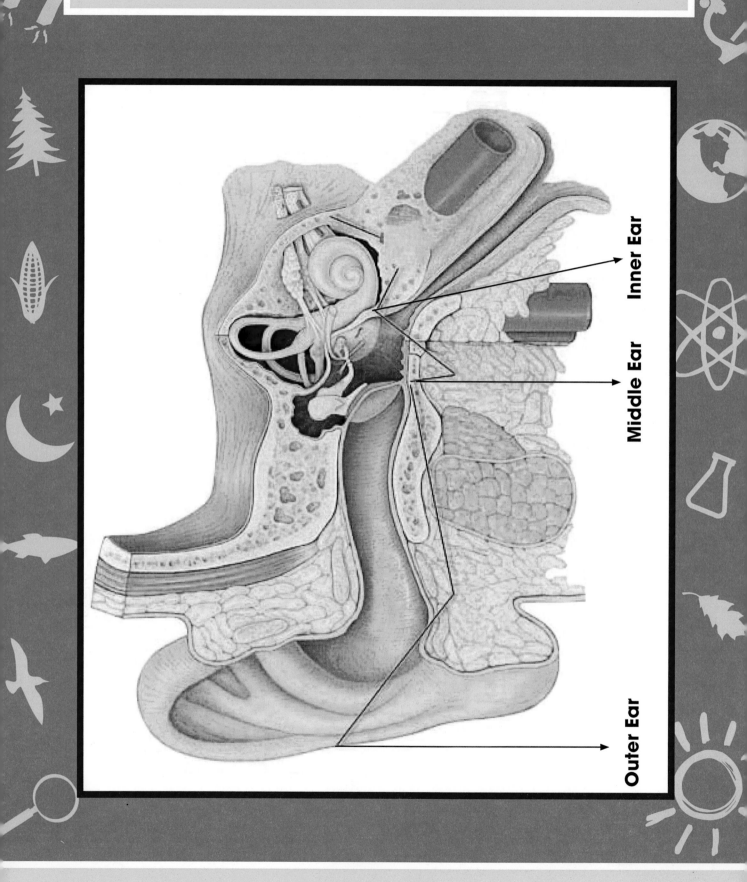

Inner Ear

Middle Ear

Outer Ear

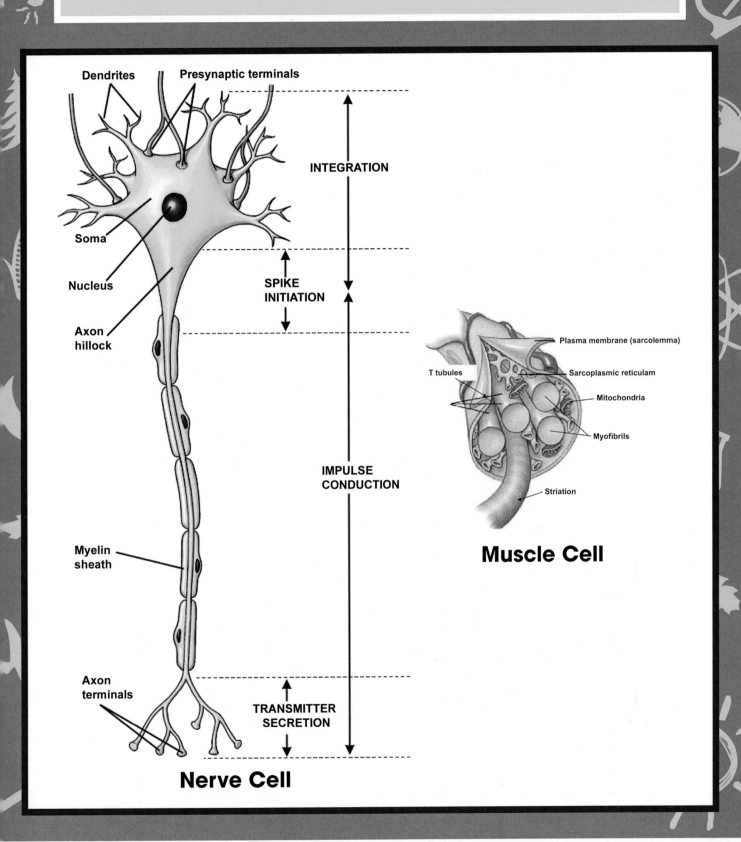

Nerve Cell

- Dendrites
- Presynaptic terminals
- Soma
- Nucleus
- Axon hillock
- INTEGRATION
- SPIKE INITIATION
- IMPULSE CONDUCTION
- Myelin sheath
- Axon terminals
- TRANSMITTER SECRETION

Nerve Cell

Muscle Cell

- Plasma membrane (sarcolemma)
- T tubules
- Sarcoplasmic reticulam
- Mitochondria
- Myofibrils
- Striation

Muscle Cell

Skeletal System

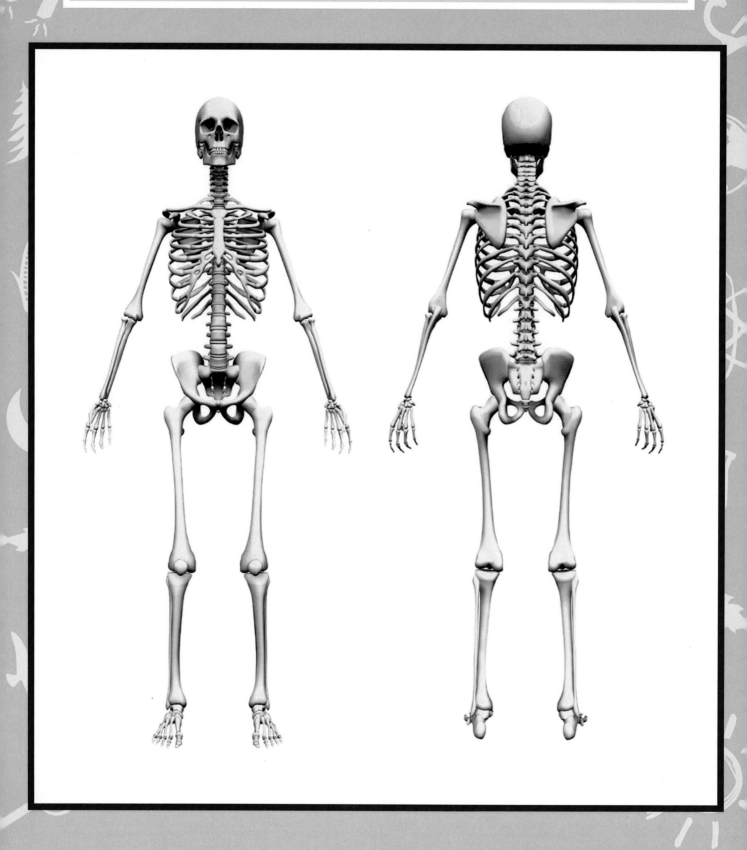

Muscles with Skull

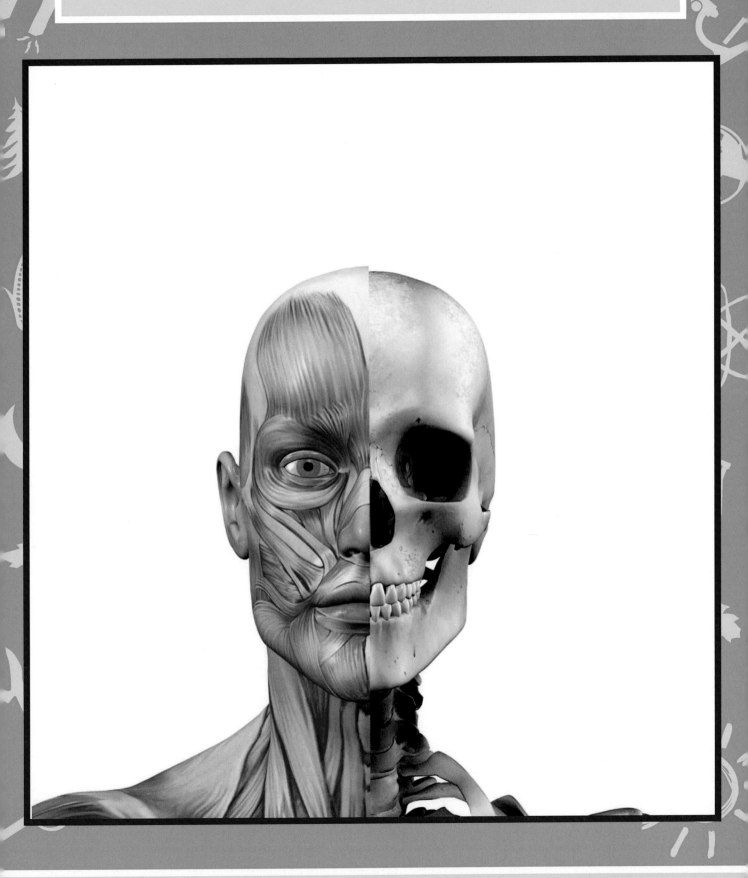

Skeletal Runner

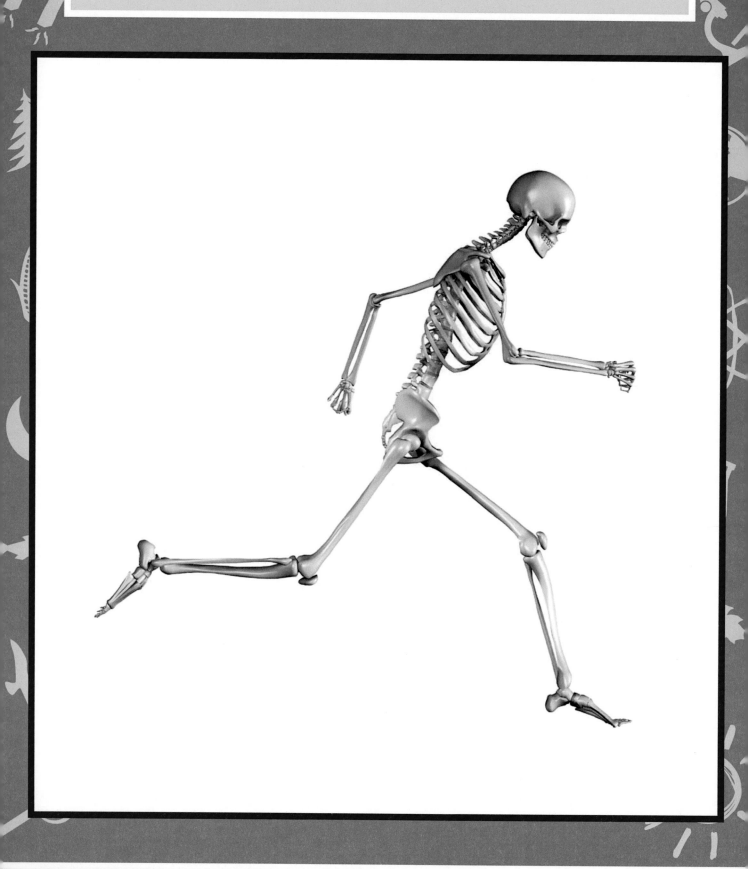

156

Skeletal-Muscular Runner

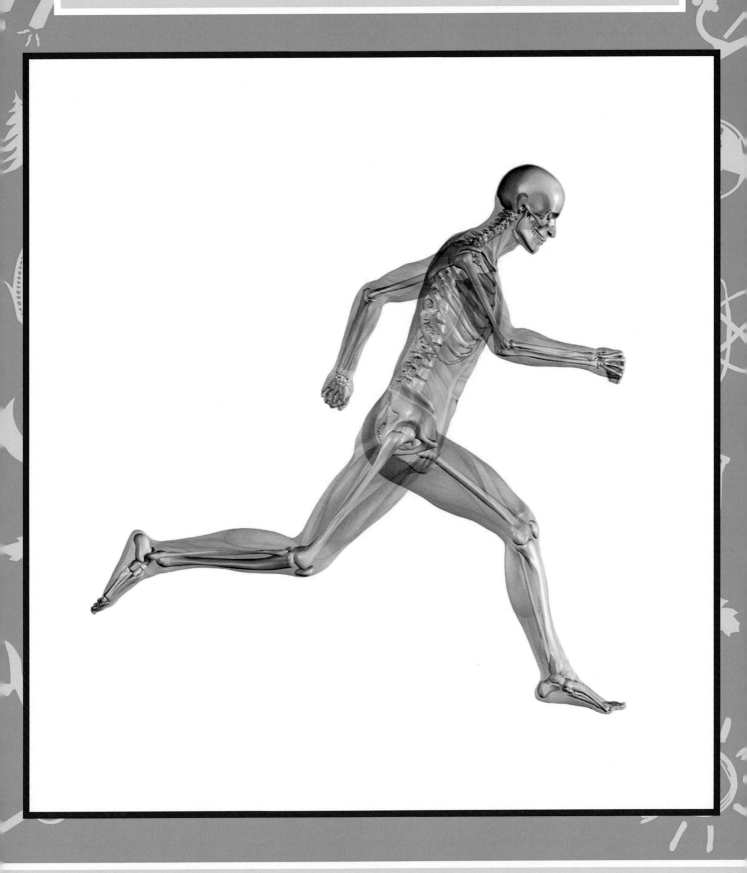

Muscular Runner

Cells, Tissues, Organs and Systems

How do tiny cells come together to form animals and plants? In most multicellular organisms, whether it is a cat, an elephant or a human, smaller (**simple**) parts join together to create bigger (**complex**) parts.

A way to understand this is to think of the parts of a sandwich. It takes bread, cheese, cold cuts and lettuce to make a sandwich. The simple things (bread, cheese, meat and lettuce) are put together to make the more complex thing (your lunch!). The chart below shows you, step-by-step, how multicellular organisms are organized from simple cells to form complex organisms.

Cellular Level   All living things are made of **cells**.

Tissue Level   **Cells** group together to form **tissues**.

Organ Level   **Tissues** group together to form **organs**.

Organ System Level   **Organs** group together to form **organ systems**.

Whole Organism   **Organ systems** group together to form **whole organisms**.

1. Which is SIMPLER? Circle your answers.

a) organ   organism   b) tissue   cell

2. Which is more COMPLEX? Circle your answers.

a) tissue   organ   b) organ system   whole organism

📖 Reading Passage

Cells, Tissues, Organs and Systems

T **issue** is a group of cells that work together in the organism to do a specialized job. There are **four main tissue types** in the human body.

The chart below shows what each of these types of tissue does and where in the body they are found.

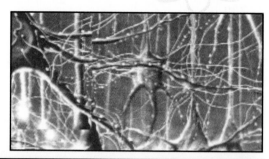

Tissue Type	What It Does	Examples
epithelial tissue	• Covers and lines the surfaces of major organs • Helps keep these organs separate, in place and protected	• Outer layer of skin • Inside lining of the digestive system
connective tissue	• Gives support and structure to the body	• Inner layers of skin, tendons and ligaments • Blood is a type of connective tissue
muscle tissue	• Specialized tissue that can change size by **contracting** (shortening) and **flexing** (lengthening) • Allows the body to move	• Smooth muscles (inside lining of organs) • Skeletal muscles (attached to bones) • Cardiac muscles (in the heart)
nerve tissue	• Carries messages, in the form of electrical signals, through the body	• Brain • Spinal cord • Nerves

1. What do you think MUSCLE TISSUE is made of?

_____

2. What do you think NERVE TISSUE is made of?

_____

Cells, Tissues, Organs and Systems

1. Fill in each blank with a term from the list.

tendons	**skin**	**epithelial**	**tissue**
blood	**brain**	**nerve**	**connective**
muscle	**spinal cord**	**smooth**	

There are four main types of _____ in the human body. _____ tissue
                                           a   **b**

carries important electrical signals around our body. Our _____ and _____ are
                                                              c   **d**

examples. _____ tissue adds support to the body. _____ are an example,
                    e   **f**

as is _____ . _____ tissue allows the body to move. One example of this type
          g   **h**

of tissue is _____ muscles. _____ tissue lines and covers our organs. The
                  i   **j**

outer layer of our _____ is an example of this type of tissue.
                          k

2. ⬭Circle⬭ **T** if the statement is **TRUE** or **F** if it is **FALSE.**

T F a) Organisms are simpler and less complex than organs.

T F b) Organ systems group together to create whole organisms.

T F c) Epithelial tissue covers the surface of our major organs.

T F d) There are five main types of tissue in the human body.

T F e) Tendons and ligaments are examples of connective tissue.

T F f) All living things are made of cells.

T F g) Cells are smaller and simpler than either tissues or organs.

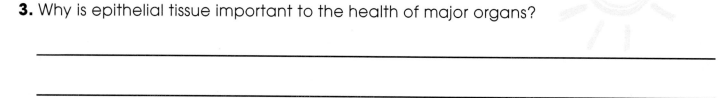

NAME: _____

Cells, Tissues, Organs and Systems

3. Why is epithelial tissue important to the health of major organs?

_____

_____

4. What is the difference between simple parts and complex parts of an organism?

_____

_____

_____

5. In the flow chart below, list these five levels in order from simplest to most complex: **organism, organ, tissue, cell, organ system**

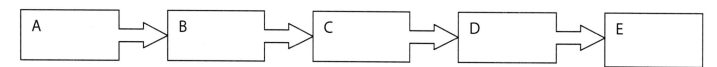

| A | B | C | D | E |

SIMPLEST   **MOST COMPLEX**

Extension & Application

6. Human body tissues and plant tissues have **similarities** and **differences**. Investigate both. Record your findings in a Venn diagram like the one below, comparing and contrasting each.

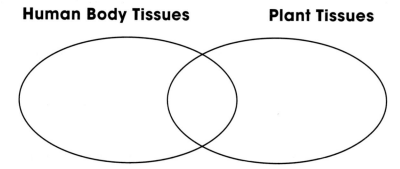

Human Body Tissues   **Plant Tissues**

7. We have read about the ideas of "simple" and "complex". These ideas apply to many different things. Make a list of all the different meals you like to eat at dinnertime. What are the **simple ingredients** that make up the **different dishes**? What dishes make up the **meal**? Record your ideas in a chart that shows how simple ingredients are put together to make dishes, and dishes are put together to make a complex meal.

NAME: _____

What are Organs and Organ Systems?

1. **Use the terms in the box to answer each question. Some terms will be left over.**

carbon dioxide	skin	bones	lungs	brain
waste	nutrients	blood	eyes	

_____ **a)** This organ is an important part of the nervous system because it sends messages to all other parts of the body. What is it?

_____ **b)** What is the largest organ in the human body?

_____ **c)** These have the important job of exchanging gases in our body. What are they?

_____ **d)** What makes up the skeletal system?

_____ **e)** Our excretory system removes this from our body in order to keep us alive. What is it?

2. In the chart below, list **at least one** way that each body system has helped you today. Once you are done, compare your answers with those of a classmate.

Organ System	How this System has Helped Me Today
Skeletal	
Muscular	
Circulatory	
Nervous	
Respiratory	
Digestive	
Excretory	

What are Organs and Organ Systems?

3. How do groups of tissues become organs?

_____

_____

4. How many major organ systems are there in the human body?

_____

_____

_____

5. a) <u>Underline</u> the terms that relate to the nervous system. (Circle) the terms that relate to the respiratory system. Be careful! Not all the terms apply.

muscles	oxygen	sexual reproduction	brain	spinal cord
lungs	bone	electrical messages	carbon dioxide	food
breath	nerves			

b) <u>Underline</u> the terms that relate to the excretory system. (Circle) the terms that relate to the circulatory system. Be careful! Not all the words apply, and one word applies to *both systems.* **Put a box** around this term.

bones	brain	waste	blood	carbon dioxide	lungs
oxygen	heart	skeleton	nutrients	sexual reproduction	

Extension & Application

6. Using the information from this lesson and your own research, decide which organ should get the **"Most Valuable Organ" Award**. Create a poster celebrating this award and include all the reasons why this organ is the winner. Make sure to include a picture of the winning organ, too!

NAME: _____

The Skeletal System – Bones

1. Use a dictionary to look up the meanings of the words below. Write the definitions in the space beside each word.

compact	
marrow	
support	
spongy	
protection	
layer	

2. Complete each sentence with a word from the list. Use a dictionary to help you.

bones	skeleton	skull	ribs	marrow	calcium

a) _____ is the mineral in bones that makes them strong.

b) Our system of bones is called our _____.

c) Red blood cells are created deep inside our bones in the

_____.

d) You cannot *see* your _____ because they are under your

skin – but you can *feel* them.

e) Our _____ is the bone that protects our brain.

f) Important organs like our heart and lungs are kept safe behind our

_____.

The Skeletal System – Bones

All of the bones in your body make up your **skeleton**. Your skeleton is also called the **skeletal system.**

1. Write TWO things you know about bones or your skeleton.

_____

2. Write a QUESTION that you have about bones or your skeleton.

_____

You have probably never seen any of your bones, but you can feel their hardness through your skin. It is this hardness that allows bones to do two important jobs: to give **support** and to give **protection**. The skeleton gives the body structure and support like the veins in an umbrella. Without bones humans would not be able to run, stand or even move. Bones also protect other parts of our body. For example, the brain is protected by the skull; the liver, heart and lungs are covered by the ribs; and the spinal cord is inside the back bones.

Red marrow in spongy bone

Hyalin(articular) cartilage

Epiphysis (bone end)

Remnant of epiphyseal plate

Diaphysis (bone shaft)

Marrow (medullary) cavity

Yellow marrow

Compact bone

Periosteum

The Parts of a Bone

All bones are made of living cells, blood **vessels** and nerves. They usually have *three layers:*

Compact Bone: The first, outside layer is thin, strong and hard bone. This compact bone contains calcium which makes it tough.

Spongy Bone: The second, middle layer has many tiny holes and looks like "cleaning sponge". The holes allow spongy bone to be strong, but not too heavy.

Bone Marrow: The third, inside layer is made of a soft jelly-like substance called bone marrow. This is where red blood cells are made.

3. Write ONE thing you have learned about bones or the skeleton. Write your answer in a complete sentence.

_____

_____

The Skeletal System - Bones

1. Put a check mark (✓) next to the answer that is most correct.

a) Why is it important that bones are hard?

   ○ **A** they give our hair support

   ○ **B** they keep our muscles strong

   ○ **C** they give us support and protection

   ○ **D** hard bones make it easier to make red blood cells

b) Bone marrow is the third inside layer of bone. What does bone marrow do?

   ○ **A** gives our body strength

   ○ **B** protects the rest of the bone

   ○ **C** creates red blood cells

   ○ **D** creates calcium

c) What is the skeletal system made of?

   ○ **A** bones and muscles

   ○ **B** bones

   ○ **C** bones and veins

   ○ **D** bones and the brain

2. Here is a picture of a bone. It shows each of the **three layers** that we have learned about. **Label** each layer using the terms below.

bone marrow   **compact bone**   **spongy bone**

a) _____

b) _____

c) _____

The Skeletal System - Bones

3. What are the three layers of bone?

_____

_____

_____

4. The middle layer of our bones is light. What would happen if it was heavy instead?

_____

_____

5. What are three organs that are protected by the ribs? Can you think of a fourth organ that is also protected by the ribs?

_____

_____

Extension & Application

6. Humans are **protected** by their skeleton. Turtles are protected by a hard shell. In this way, the shell does the same job as a skeleton. Do some research to learn more about a turtle's shell and the human skeleton. Then, **compare** the skeleton and a turtle's shell. How are they similar and how are they different? Here are some questions for you to consider as you collect your facts:

- **What does each look like?**
- **Where is each located in the body (i.e., on the inside or on the outside)?**
- **What other job or jobs, besides protection, does each have?**
- **How do they grow?**
- **What do they need to stay healthy?**

Present your findings as a one-page report. Add pictures if you like.

7. Do some research to find out about **three diseases** that affect the bones in the human body. Then, look for information on ways to keep our bones **healthy**. Copy the T-chart below into your notebook. Complete the chart with the information you collect. Share your findings with a classmate.

3 Diseases of the Bones	3 Ways to Keep the Bones Healthy

The Skeletal System – Joints and Cartilage

1. **Circle** **the word that completes the sentence. You may use a dictionary to help.**

a) The bones of the skeletal system are held together by _____.

joints   skin

b) The elbow and knee are examples of _____ joints because they can swing forward and backwards, just like a door.

handle   hinge

c) A ball and _____ joint is called this because the ball of one bone fits into the hollow area of the other bone.

soccer   socket

d) Our wrists can turn in a complete circle, moving in all directions. This is called _____.

relocation   rotation

e) The ends of our bones are protected by a rubbery material called _____.

cytoplasm   cartilage

f) The ends of our bones need protection because otherwise they would wear down from _____ on each other.

banging   grinding

2. **In the chart below, list what you already know about the skeletal system and some questions you have about it.**

What I Know About the Skeletal System	Questions I Have About the Skeletal System

The Skeletal System – Joints and Cartilage

What Are Joints?

Your skeletal system is made up of 206 different bones. Bones are connected to each other by **joints.** Without joints, bones would not be able to move because it is at the joint that movement takes place. Three of the most important types of joints are the ball and socket, hinge, and sliding joints.

1. Ball and Socket Joint: This kind of joint allows for movement in *almost any direction,* like a computer joystick. Ball and socket joints are found in the shoulder and the hip.

○ Nonaxial
◒ Uniaxial
◐ Biaxial
○ Multiaxial
○ Ball and socket joint

2. Hinge Joint: This type of joint allows for *forward* and *backward* movement, like the hinge of a door. Elbows and knees are hinge joints. A hinge joint does not allow for as much movement as a ball and socket joint, but it is stronger.

3. Sliding Joint: This type of joint lets bones *slide* easily across each other. This allows both bending and turning (rotation). Ankles and wrists have sliding joints.

Think of all the places in your body where bones join together to form joints. Besides elbows and knees, what is another joint that might be a HINGE JOINT? (Remember how a hinge joint moves...)

_____

_____

What Is Cartilage?

The ends of many bones are covered with a tough rubbery material called **cartilage**. One of the main jobs of cartilage is to protect bones at the joint. Without cartilage, bones would grind against each other when we moved them. In time the bones would wear away. Besides our joints, did you know that our ears and the tip of our nose are cartilage, too?

Here is another interesting fact about cartilage: Most of the twenty-nine bones in your skull are held together by joints made of cartilage. These joints can move a bit in babies, but by the time we are fully grown they do not move at all.

The Skeletal System – Joints and Cartilage

1. **Circle** **T** if the statement is **TRUE** or **F** if it is **FALSE**.

   T F **a)** The human skeleton is made of 406 bones.

   T F **b)** The main job of our joints is to help bones move.

   T F **c)** The hip is an example of a hinge joint.

   T F **d)** The wrist is an example of a hinge joint.

   T F **e)** Cartilage protects bones from grinding against each other when they move.

   T F **f)** The shoulder is an example of a ball and socket joint.

   T F **g)** A hinge joint is stronger than a ball and socket joint.

   T F **h)** There are 29 bones in our skull all joined by cartilage.

   T F **i)** Elbows and knees are both examples of hinge joints.

   T F **j)** Sliding joints help wrists and ankles move in all directions.

2. **Here are pictures of each of the three kinds of joints that we have learned about. Label each picture with the correct joint name.**

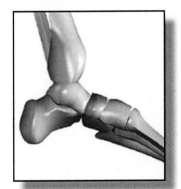

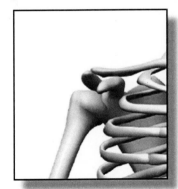

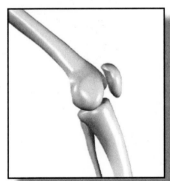

a) _____

_____

b) _____

_____

c) _____

_____

The Skeletal System – Joints and Cartilage

3. Why are joints important?

_____

_____

4. How do you think walking would feel if we had no cartilage in our leg bones?

_____

_____

5. What joint do you think is the most important and why?

_____

_____

Research, Extension & Application

6. Humans get slightly **shorter** between adulthood and old age. Do some research to find out why this happens.

7. If a hip or knee joint becomes diseased it can be replaced with an **artificial joint**. Do research to find some interesting facts about this type of surgery. Think about these questions as you collect your facts:
  ● **How is a joint replacement done?**
  ● **What material is the artificial joint made from? How is it similar to and different from bone?**
  ● **What kind of doctor does this surgery?**

8. Every time you take a step your knee joints work. Use a **pedometer** to count the number of steps you take in an average day. Using this number **calculate** how many steps you take in a week (seven days), in a month (30 days) and in a year (365 days).

9. Calcium is a mineral important to having strong bones. Do some research to find **ten foods** high in calcium.

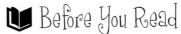
The Muscular System – Muscles

1. **Match the words on the left to the definitions on the right. You may use a dictionary to help.**

bundled	**A**	A slender, threadlike object
cardiac	**B**	A group of things gathered close together are _____
involuntary	**C**	Another word for striped
striated	**D**	A word used to describe things related to the heart
fiber	**E**	Something that happens without thinking about it or choosing it to happen is _____

2. The muscular system has two main jobs. Can you guess what they are? Think about how your body works and what you already know about muscles. Then <u>underline</u> your two guesses from the list below.

muscles help us think faster
muscles give our body shape
muscles help us move

muscles help us sleep better
muscles give us better hearing

3. In the chart below list some of the things your muscles have helped you to do today. Try to list at least five different things.

1.

2.

3.

4.

5.

Thanks, muscles! Without you I wouldn't have been able to do any of these things today!

The Muscular System - Muscles

There are over 600 muscles in the human body. Together they make up the **muscular system.**

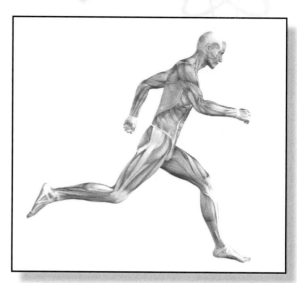

Just like your skeleton, your muscles are below the skin. This means that you can only see the outline of muscles and feel them change shape when they move. Try this – hold your arm up like a weightlifter and tighten your fist. Can you feel the muscles in your upper arm getting bigger and harder? Doing this exercise gives you a clue about the two main jobs of muscles:

1. Muscles give our body shape; 2. Muscles help us move. Muscles help us do almost everything – from running on the playground, to pulling the blankets up at night, to breathing in and out.

All muscles in the body are made of cells that are like elastic string. These cells are called **muscle fibers**. Muscle fibers are **bundled** together in groups to form muscles. The number and length of the fibers depends on the size of the muscle.

Types of Muscle

There are three types of muscle in the human body. The chart below tells you about them.

Muscle Type	What It Looks Like	How It Moves	What It Does
Skeletal	Striated (striped)	Voluntary (we can control the movement)	Attached to our bones, allowing them to move
Smooth	Smooth	Involuntary (we cannot control the movement)	Controls movement inside our body
Cardiac	Striated (striped)	Involuntary (we cannot control the movement)	Allows our heart to pump blood

STOP

What surprising thing have you learned about muscles? Why did it surprise you?

_____

_____

The Muscular System – Muscles

1. (Circle) **T** if the statement is TRUE or **F** if it is FALSE.

T	F	**a)**	The human body has over 1000 muscles.
T	F	**b)**	Muscles help us move.
T	F	**c)**	Cardiac muscle is voluntary muscle.
T	F	**d)**	Every human muscle is made of one large muscle fiber.
T	F	**e)**	If something is "involuntary" it means we cannot control it.
T	F	**f)**	The human body has three important muscles on top of the skin.
T	F	**g)**	Muscles give our body shape.
T	F	**h)**	Smooth muscles are striated.
T	F	**i)**	Cardiac muscles are involuntary muscles that help pump our blood.

2. Fill in each blank with a word from the list. There will be two words left over.

bundled	shape	three	cells	run	two	fibers
breathe	move	elastic	muscles			

Muscles have _____ main jobs. They give our body _____
                     a   **b**

and help us to _____. Without muscles we couldn't _____
                        c   **d**

or _____. All muscles are made of _____ like
             e   **f**

_____ string. These are called muscle _____. These fibers
            g   **h**

are _____ together to create muscles.
            i

     Human Body CC4519

The Muscular System - Muscles

3. What are two ways muscles help us?

_____

_____

4. What do skeletal and cardiac muscles look like?

_____

_____

5. What muscle do you think is the most important and why?

_____

_____

Extension and Application

6. Below is a list of words from the reading. Choose five words that link together well. Then write a **short story** about a day in the life of a muscle. Be imaginative!

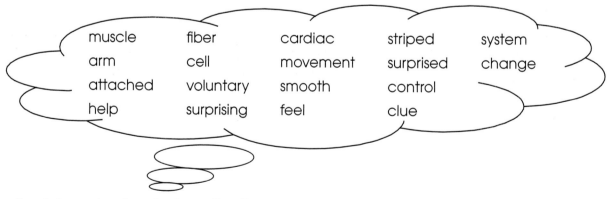

muscle	fiber	cardiac	striped	system
arm	cell	movement	surprised	change
attached	voluntary	smooth	control	
help	surprising	feel	clue	

7. Using the Internet or books from the library, research **six** different foods that help build healthy muscles. Create a menu, including these foods, to promote muscle-healthy eating.

8. **Multiple Sclerosis** is a serious muscle disease. Research important facts about this disease in an encyclopedia or on the Internet. Write a three-paragraph report based on the information you find. Try to find out when it was discovered. List its symptoms and treatments.

The Muscular System – Movement

1. Use a dictionary to look up the meanings of the words below. Write the definitions in the space beside each word.

voluntary	
tendon	
involuntary	
strain	
esophagus	

2. The body can do each of the things listed below. Decide if each action is **voluntary** or **involuntary.** Here's a hint if you get stuck: Ask yourself if you can choose to do this thing, or if your body does it without you deciding. Write your answers in the chart.

biting an apple	kicking a ball	your heart beating
blood moving in your veins	getting goose bumps at a scary movie	digesting an apple
walking to the bus		

Voluntary	Involuntary

The Muscular System – Movement

We have read that muscles and bones are both needed for movement. To work together to create movement, muscles and bones have to be connected. Most muscles are attached to bone by strong cords called **tendons**. Tendons look like rubber bands. Besides connecting bone and muscle, one other important job of tendons is to protect muscle from strain during movement.

Can you find any tendons in your body? Try this: Touch the top of your hand as you wiggle your fingers. The hard ridges you feel that run from your fingers to your wrist are tendons.

Upper Arm Muscles

Triceps tendon

Biceps tendon

Triceps

Biceps

Back

Front

Involuntary Muscle Movement:

There are two ways that muscles can move. One kind is **involuntary movement. Smooth** muscles and **cardiac** muscles both move involuntarily. This means they move on their own; we cannot decide to *make* them move. Our brain sends messages to these muscles "telling" them when they need to move. This happens without us even knowing our muscles are working. An example of involuntary movement is in our digestive system. When we swallow food, the food is pushed down into our stomach by rings of smooth muscles in our **esophagus.** Our stomach is lined with smooth muscle, too. This muscle moves around food we have eaten, breaking it up into small bits so we can get the nutrients from it.

STOP

What are TWO kinds of muscle that move INVOLUNTARILY?

_____

_____

The Muscular System – Movement

One of the most important involuntary muscles is our heart. It is made of cardiac muscle. Cardiac muscle beats, or **contracts,** through our whole life, sending blood to every part of our body. On average, the heart beats 70 times a minute. By the time you are 70 years old your heart will have beaten two and a half billion times! Can you imagine if you had to remember to make your heart beat? It would be impossible.

Voluntary Muscle Movement

The second kind of muscle movement is **voluntary.** Voluntary movement happens when we choose to move our muscles; we can control this kind of movement. Three important body parts come together allowing us to move. These are our brain, muscles and bones.

Let's look at an example of voluntary muscle movement: It's lunch time and you're hungry. You want to take a big bite of your turkey sandwich. But how does your body get your arm to move to your mouth? The thought, "I'm going to take a bite of my sandwich" causes your brain to send an electrical message through your nerves to two muscles (a pair) in your arm. This **muscle pair** works together to move your arm. One muscle in the front of your arm *shortens* (contracts). At the same time the muscle in the back of your arm *lengthens.* This pulls the bone in your lower arm and raises it up towards your mouth. Most skeletal movement happens this way, thanks to the connection of our brain, muscles and bones.

Brain, Muscles and Bones Create Voluntary Movement

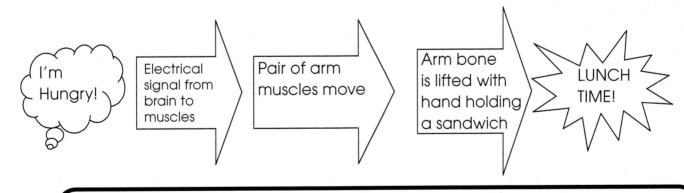

Why is eating a sandwich an example of VOLUNTARY movement?

_____

_____

The Muscular System – Movement

1. Fill in each blank with a word from the list. There will be four words left over.

muscle	arm	voluntary	involuntary	smooth	move
heart	digestive	brain	can	cannot	

Cardiac and _____ muscles are both types of
          a

_____ muscles. We _____ control their
    b   **c**

movement. Our _____ sends signals to these muscles to
                         d

make them _____, and we don't even know it's happening.
                 e

The muscles of our _____ system are involuntary muscles. To make
                     f

_____ muscles move we have to decide to move them. This is an
    g

example of voluntary muscles.

2. (Circle) **T** if the statement is **TRUE** or **F** if it is **FALSE**.

T   F   **a)**   Most muscles are attached to bone by a strong cord called ligament.

T   F   **b)**   Three important body parts come together in voluntary movement.

T   F   **c)**   These three important body parts are the brain, muscles and internal organs.

T   F   **d)**   When a muscle shortens it contracts.

T   F   **e)**   Involuntary muscles work in pairs to move.

T   F   **f)**   An example of tendons are the hard ridges that run from our fingers to our wrist.

T   F   **g)**   Smooth muscles and cardiac muscles are both involuntary.

T   F   **h)**   Tendons help protect muscles from strain.

The Muscular System – Movement

3. Why do you think it is important that the heart is an involuntary muscle?

_____

_____

4. Why do you think it is important that the muscles in our legs are voluntary muscles?

_____

_____

5. How do the muscles in your arm work together to create movement?

_____

_____

_____

Extension & Application

6. Using the Internet, research **ten** different **world records** of human movement. Some examples include:

fastest 100 meters   **fastest marathon**   **highest pole vault**

highest high jump   **fastest 100 meter speed skate**

fastest 500 meter breast stroke (swimming)

For each world record, write down the name of the winner, whether it is a record for men or women, and what country the winner is from.

Record your findings in a chart like the one below.

World Record	Name of Winner	Male or Female?	Country

Most of these world records will be for voluntary muscle movement. Can you find any that are for **involuntary movement?**

Build Your Own Cell

We have learned that while cells can come in different shapes and sizes, they all have some parts in common. Do you remember what these are?

1. The **cell membrane** is the outside covering that separates inside of the cell from its environment.

2. The **cytoplasm** is the jelly-like substance inside the cell where all the work takes place.

3. The **nucleus** floats in the cytoplasm and contains DNA.

4. The **mitochondria** float in the cytoplasm too, and turn food into energy.

5. The **lysosomes** also float in the cytoplasm, and keep the cell clean.

FOR THIS ACTIVITY, you will need:

- 5 different colors of plasticine
- 5 toothpicks
- small pieces of paper
- tape

STEPS:

1. Use plasticine to **sculpt** your cell. First, decide what shape it will be. Remember that human body cells can be long and thin, round, or rectangular in shape. Use a different **color** for each cell part. The cell should be **at least** the size of your hand.

2. Once you have finished sculpting your cell, place the toothpicks in the plasticine. You will use them as markers for the different cell parts.

3. On small pieces of paper, write down the cell parts. 'Flag' them by taping them to the toothpicks.

When you are finished, someone should be able to look at your plasticine cell and to see the five different parts properly labeled. Have fun sculpting!

Create a Human Body Organ System Booklet

We have learned that the human body has EIGHT major organ systems. Each system is made up of important ORGANS, and these organs work together as a SYSTEM. All of these organ systems have important jobs to do to keep our body healthy and alive.

Your task is to create a booklet with important facts about each of the organ systems:

skeletal system	muscular system
circulatory system	nervous system
respiratory system	digestive system
excretory system	reproductive system

YOUR BOOKLET SHOULD INCLUDE:

- a cover page with the title of your booklet
- a table of contents page
- at least one page for each organ system

COLLECTING YOUR INFORMATION:

Begin by collecting important **facts** about each system. You may use the reading passages, the Internet, and other resource materials to find your information. For each organ system, include the following information:

1. **Major organs** that make up the system

2. The **main jobs** of the organ system (what it does)

3. A **picture** that shows what the system looks like (be sure to label all the parts!)

4. Other interesting facts that you find

Invent an Alien Skeleton!

This activity has two parts. In the first part, you will label the bones in the human skeleton. In the second part, you will use what you have learned to invent your own extra-terrestrial skeleton.

Part 1

Use the words in the list to **label** the bones on the skeleton. You may need to do some research to complete this part.

ribs
patella
backbone
pelvis
femur
tibia
scapula
sternum
humerus
fibula
clavicle
radius
phalanges
ulna
mandible

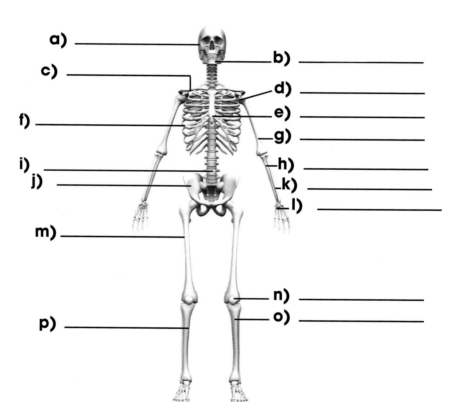

a) _____
b) _____
c) _____
d) _____
e) _____
f) _____
g) _____
h) _____
i) _____
j) _____
k) _____
l) _____
m) _____
n) _____
o) _____
p) _____

Part 2

Now, it is time to **draw** your own **alien skeleton!** You must use **at least ten** different skeleton parts from the diagram above. You may use the parts more than once if you like. Be as imaginative as you can! Draw your skeleton on a separate sheet of paper. Above your drawing, copy and complete the following:

Hello! I am an extra-terrestrial from the planet _____. My name is _____ and my favorite food is _____. I have _____ skulls, _____ femurs, _____ tibias, and _____ ribs.

Pin the Organ on the Body

Below is an outline of the human body. On the left side of the page are pictures of important **ORGANS** in the body. Your task is to **CUT OUT** each organ and to **PASTE** it on the body where it belongs. You may use information from the reading passages, the Internet, or other resource materials to find the answers.

a) liver

b) intestines

c) brain

d) bladder

e) heart

f) lungs

g) stomach

h) kidneys

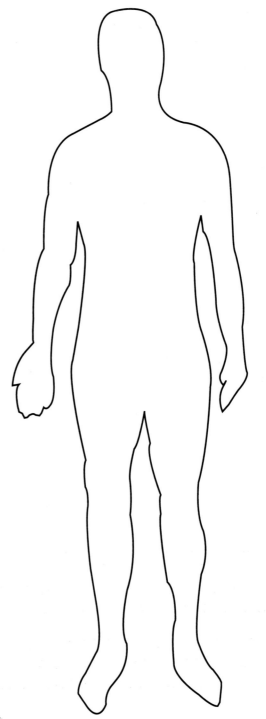

Human Body CC4519

Crossword Puzzle!

Across

1. Muscle tissue changes size by _____ and lengthening
3. The human body is made of _____ cells
4. Muscle_____ are like elastic string
6. Humans are_____ organisms
8. The knee is an example of a _____ joint
10. _____ muscles allow our bones to move
11. Cells group together to form _____
13. Cells contain special information called _____
14. There are _____ major organ systems in the human body

Down

1. The liquid inside a cell is called _____
2. The skeletal system is made of bones, joints and _____
4. The heart is made of _____ muscle
5. Nerve tissue carries messages from the brain in the form of electrical _____
7. The digestive system is made of mostly _____ muscles
9. Muscles work in _____; one shortens and the other lengthens
12. Mitochondria turn food into _____

Word List

specialized	cytoplasm
contracting	energy
DNA	cardiac
cells	involuntary
tissues	pairs
hinge	cartilage
multicellular	signals
eight	skeletal

Word Search

Find all of the words in the Word Search. Words are written horizontally, vertically, diagonally, and some are even written backwards.

multicellular
tissue
calcium
specialized
unicellular
marrow
nucleus
organ
compact
cytoplasm
muscle
cartilage
mitochondria
nerve
tendon
lysosomes
skeletal
contract
complex
cardiac
socket
joint
involuntary
hinge
fiber
striated
bundled
rotation

Q	A	Z	X	S	M	C	T	C	A	P	M	O	C	S	W	E	U
P	L	K	S	A	A	D	K	G	O	P	L	M	K	P	S	N	E
W	S	D	R	I	E	D	S	U	E	L	C	U	N	E	I	J	B
S	Y	R	D	H	N	M	J	C	O	N	T	R	A	C	T	K	U
C	O	R	T	G	B	Y	H	N	M	J	U	K	E	I	P	L	M
W	A	C	X	D	S	K	E	L	E	T	A	L	D	A	R	F	X
C	T	L	K	Y	H	N	U	J	M	K	L	E	D	L	T	J	K
B	G	T	C	E	Y	H	N	U	K	U	O	P	H	I	N	G	E
E	D	T	A	I	T	Y	H	N	L	I	U	J	M	Z	I	K	P
W	S	T	R	Y	U	H	N	A	M	N	E	R	V	E	K	G	B
A	Z	D	T	F	G	M	R	H	J	V	T	H	U	D	Y	H	A
U	J	O	I	N	T	Y	L	Y	S	O	S	O	M	E	S	G	I
V	F	R	L	A	G	B	N	M	J	L	K	H	F	I	B	E	R
C	U	J	A	G	J	K	L	M	B	U	N	D	L	E	D	C	D
Y	O	H	G	R	S	M	R	I	A	N	E	D	M	J	K	U	N
T	Q	M	E	O	S	T	R	I	A	T	E	D	U	C	V	B	O
O	T	B	P	Y	E	S	R	O	T	A	T	I	O	N	M	D	H
P	D	C	T	L	T	Y	J	U	K	R	N	K	U	T	G	H	C
L	T	D	C	H	E	I	R	O	T	Y	T	I	O	N	Y	F	O
A	D	S	Z	X	C	X	S	T	G	B	N	N	O	D	N	E	T
S	U	S	D	F	T	H	U	S	D	Y	A	W	E	R	Q	F	I
M	U	L	T	I	C	E	L	L	U	L	A	R	A	Y	G	C	M
D	E	L	D	N	U	B	D	S	X	E	Y	H	N	M	K	U	P

Comprehension Quiz

32

Part A

Circle **T** if the statement is **TRUE** or **F** if it is **FALSE**.

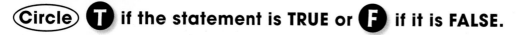

8

T F **1)** The cell nucleus contains hereditary information called DNA.

T F **2)** In the human body, organs are made of groups of tissue that have a specific job.

T F **3)** Organ systems are simpler than organs.

T F **4)** Three of the major organ systems in the human body are the respiratory system, the skeletal system and the brain system.

T F **5)** In the circulatory system, the heart pumps blood through our nerves.

T F **6)** The main jobs of the skeletal system are to give protection and support.

T F **7)** The ends of our bones are covered by a rubbery material called cytoplasm.

T F **8)** Skeletal muscles control the digestion of food in our stomach.

Part B

On the diagram below, label the three layers of bone. Use the words in the list.

6

bone marrow   spongy bone   compact bone

1. _____

2. _____

3. _____

SUBTOTAL: /14

After You Read 📖

• • • • • • • • • • • • • • • • • • • •

Comprehension Quiz

Part C

Answer each question in complete sentences.

1. What are **specialized cells?** Are they found in unicellular or multicellular organisms? Give an example of an organism that is made of specialized cells. ③

_____

_____

2. Name **two parts of a cell.** Describe the **function** of each part in the cell. ④

_____

_____

3. Name **one kind of tissue** in the human body. Describe **what it does** in the body. Give an example of this type of tissue. ③

_____

_____

4. What is the difference between **voluntary** and **involuntary** movement? Name **one** kind of muscle that moves voluntarily. Name **one** kind of muscle that moves involuntarily. ④

_____

_____

_____

5. Describe how **voluntary movement** happens. Use the words **brain, muscle pair** and **bone** in your answer. ④

_____

_____

_____

SUBTOTAL: /18

NAME: _____

The Nervous System - Brain

1. **Complete each sentence with a term from the list. Use a dictionary to help you.**

    brain   spinal cord   nerves   messages   data

    a) The _____ is like a computer that controls our body.

    b) The information that is carried along our nervous system is like computer
        _____.

    c) Important _____ about the world around us are understood in our
        brain.

    d) Our _____ is protected inside the bones of our spine.

    e) We have millions of _____ in our body that carry messages to our
        brain.

2. **Label the nervous system using the terms in the list.**

    nerves   brain   spinal cord

    a) _____

    b) _____

    c) _____

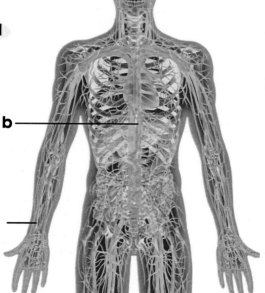

NAME: _____

The Nervous System – Brain

One of the most important organ systems in our body is the **nervous system**. The nervous system is a **network** of **tissue** that has the job of sending and carrying **messages** to all areas of our body. Our nervous system controls all our movements and reactions to the world around us. The nervous system is made of three important parts – the **brain**, the **spinal cord** and the **nerves**.

To understand how the parts of the nervous system work together, think of a **computer system**. The brain is the computer. The spinal cord is the cable carrying the messages or **data** to and from the computer. All the nerves connect to the spinal cord. These nerves carry the messages to every part of the body and also send important **information** back to the brain.

STOP

Why is the brain like a computer?

_____

_____

The Brain

The brain **manages** our nervous system. It is the control center of our body. The brain is protected inside the bones of the skull. It weighs about three pounds and is made of over 100 **billion** nerve cells. The brain has three main parts.

Brain Part	Description	What It Does
cerebrum	• the large top part of the brain • divided into two halves • looks pinkish gray in color, is jelly-like and **wrinkled**	• controls thinking, **memory**, all our **emotions** and **language** • also very important for movement
cerebellum	• also called "little brain" • sits below the cerebrum	• important for movement, **balance** and **posture**
brain stem	• looks like a **stalk** that connects the brain to the spinal cord • the simplest part of our brain.	• controls **involuntary** movements like our breathing and heart beat

The Nervous System - Brain

1. **Fill in each blank with a term from the list.**

control	cerebrum	brain	protected	brain stem
one hundred	cerebellum	three	skull	

The _____ manages our nervous system; it is the _____
         a   **b**

center of the body. The brain is _____ inside the _____
                                              c   **d**

which is made up of the bones of our head. The brain weighs _____
                                                                          e

pounds and is made of over _____ billion cells. The brain has three main
                                      f

parts: the _____ controls memory; the _____ controls
                  g   **h**

balance; and the _____ controls involuntary movements.
                          i

2. <u>Underline</u> the words and ideas that describe what the nervous system does.

control center   carries messages   controls emotion

helps us understand the world   controls breathing   controls posture

helps us understand language

3. **Use the information from the reading passage to label the parts of the brain.**

brain stem   cerebrum   cerebellum

a) _____

b) _____

c) _____

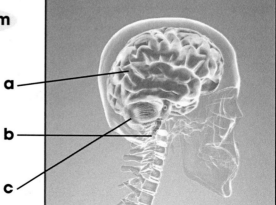

The Nervous System – Brain

4. What are **two** ways in which our nervous system is similar to a computer system?

_____

_____

5. What are the three main parts of the **nervous system**?

_____

_____

6. Which statements apply to which of the three main parts of the brain? Put an X in the correct box to the right of each statement.

Feature	Cerebrum	Cerebellum	Brain Stem
a) Controls language			
b) Also called "little brain"			
c) Helps with our balance			
d) Controls thinking and memory			
e) Split into two pinkish gray halves			
f) Controls our heart beat			

Extension & Application

7. **Alzheimer's disease** is a disease of the brain that some people get in their old age. There are other **common diseases** of the nervous system. Research Alzheimer's or another disease of the brain or of the nervous system. Look for information on the Internet and in books from the library. Record your information in point form in the **Web Organizer** on the next page. Here are some questions you should find the answers to:

- **What happens to the brain and/or nervous system?**
- **What are the symptoms?**
- **How is it treated?**
- **Are there some people who are more likely to get it than others?**
- **What can people do to avoid getting this disease (if anything)?**
- **Any other interesting facts**

NAME: _____

Disease of the Nervous System

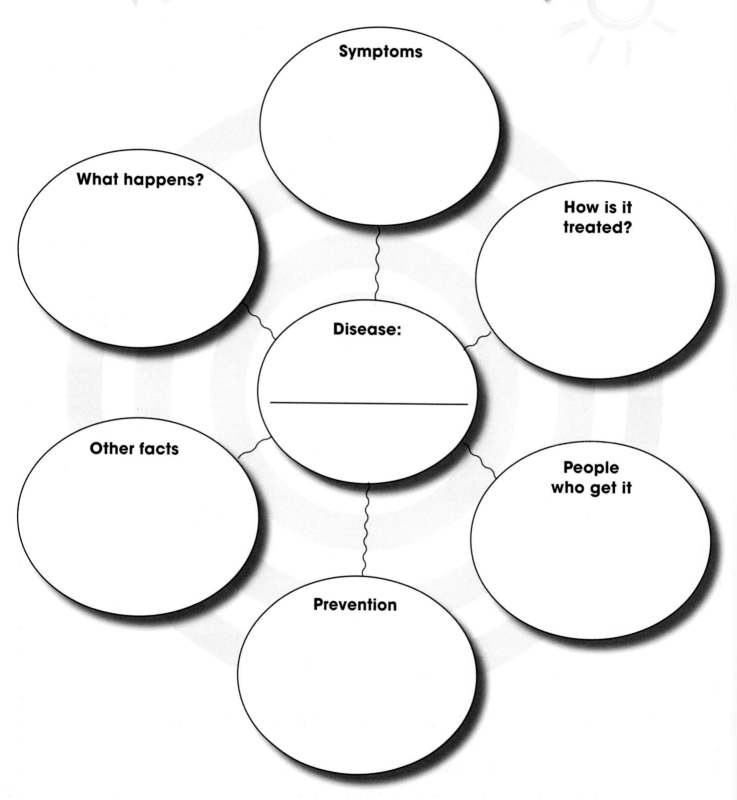

Symptoms

What happens?

How is it treated?

Disease: _____

Other facts

People who get it

Prevention

The Nervous System – Spinal Cord and Nerves

1. Complete each sentence with a term from the list. Use a dictionary to help you.

> neuron   brain   tissue   spinal cord   vertebra

a) A single bone in the spine is called a _____.

b) _____ is another word for nerve cell.

c) The _____ carries messages to and from the brain.

d) Our _____ is protected inside our skull.

e) A group of cells that work together to do a specific job are called

_____.

---

2. Which word completes the sentence? (Circle) your answer. You may need a dictionary to help you.

a) Our spinal cord is _____ by the bones of our spine.

  projected   **protected**

b) The spinal cord is a thick _____ of nerve cells that run down our back.

  bundle   **bubble**

c) Our nerve cells carry _____ to and from our brain.

  messages   **messengers**

The Nervous System – Spinal Cord and Nerves

We have read that the nervous system is like a computer network that sends messages to every part of our body. We know the brain controls the whole system. Now we'll learn about the other two important parts of the nervous system – the spinal cord and the nerves.

The Spinal Cord

The spinal cord is a thick bundle of nerve tissue that runs down our back from the base of the brain. Information travels back and forth between the brain and the nerves in the rest of the body along the spinal cord. If our spinal cord was seriously injured, this two-way communication between our body and brain would stop. This is why our spinal cord is protected *inside* the bones of the spine, called **vertebrae**.

STOP

Why is our spinal cord inside our vertebrae?

_____

_____

The Nerves

Nerve cells (also called **neurons**) have the job of carrying messages. They carry messages to and from the brain through the spinal cord to every part of the body. Nerve cells are connected to our spinal cord and to each other. They have tiny ends, like threads, so they can reach everywhere in our body.

There are two main types of nerves. The first are called **motor nerves (or motor neurons)**. These nerves work with our muscles to make movement possible. When you think, "I want to pet my dog", you can do it because your motor nerves have "talked" to your muscles. They have **communicated** the message.

The second kind of nerves are called **sensory nerves** (or **sensory neurons**). These nerves bring messages from our **senses** – ears (hearing), eyes (seeing), tongue (tasting), nose (smelling), and skin (touching) – to our brain. Without these nerves you would not be able to feel the soft fur of a kitten or hear your favorite music.

The Nervous System – Spinal Cord and Nerves

1. Put a check mark (🕐) next to the answer that is most correct.

   a) **What is the main job of the motor neurons?**

     ○ **A**   bring messages to our senses

     ○ **B**   make movement possible

     ○ **C**   protect the spinal cord from damage

     ○ **D**   control our emotions

   b) **Which best describes the spinal cord?**

     ○ **A**   a long boney ridge

     ○ **B**   a thick bundle of nerves

     ○ **C**   a computer

     ○ **D**   protected by our skull

2. Fill in each blank with a term from the list. Be careful! <u>One</u> term will be used twice. <u>Two</u> terms will be left over.

smell	brain	body	sensory	neurons	hear
three	spinal cord	two	messages	motor	

Nerve cells are also called _____. It is their job to carry _____ between the
                           a   **b**

_____ and the _____. The messages travel along our _____.
   c   **d**   **e**

Our nervous system is made of _____ kinds of neurons. _____ neurons
                                 f   **g**

make movement possible. _____ neurons carry messages from our senses.
                           h

Without these we would not be able to _____ drums or _____ pizza.
                                      i   **j**

The Nervous System – Spinal Cord and Nerves

3. Why is it important that our spinal cord is protected?

_____

_____

4. List **three** activities you wouldn't be able to do if you didn't have the use of your motor nerves.

_____

_____

5. Below are some words from the reading. Write each word beside its meaning.

neuron	nerve	vertebrae	senses	protect

_____ **a)** the bones of the spine

_____ **b)** _____ cells carry messages

_____ **c)** to keep something safe

_____ **d)** seeing and hearing are examples of these

_____ **e)** another word for "nerve cell"

Extension & Application

6. Christopher Reeve was a famous actor who became **paralyzed** when his spine was injured in an accident. Do some research to find out more about Christopher Reeve. Try to find out how his spine was injured and how his life changed when he was paralyzed. Also find out how he helped other people with spinal injuries.

7. Sledge hockey is a popular sport played at the **Paralympics**. Only people with damaged spines can play this sport. Research sledge hockey to find out how it is played at the Paralympics. Then, compare it to the way professional hockey is played. Record your information in a Venn Diagram. What is unique to each game? What is the same in both?

The Sense of Sight

1. **Match the word on the left to the definition on the right. You may use a dictionary to help you.**

eyelid	To keep something safe from harm	A
protect	The small black hole in the center of the eye	B
blink	We do this when we open and close our eyes quickly	C
pupil	Thin layer of skin that slides down over the eye	D

2. **In the chart below, list your** (top ten) **favorite sights of all time. These could be things you see every day. Or, they may be things you love, but only see sometimes. Or, you may have only seen them once in your life.**

MY TOP TEN FAVORITE SIGHTS OF ALL TIME!
1. _____
2. _____
3. _____
4. _____
5. _____
6. _____
7. _____
8. _____
9. _____
10. _____

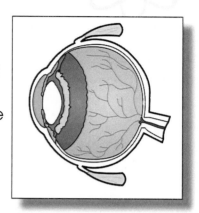

The Sense of Sight

For many people, eyesight is one of the most important of the five senses. The information we get through our eyes helps us in many ways. Seeing a car driving toward us helps us know to move out of the way. Reading a book teaches us new things about the world. Watching a funny movie entertains us and makes us laugh. We are happy when we see the faces of the people we love.

Let's look at the important parts of the eye and how they work to make seeing possible.

Eye Part	Description	What It Does
iris	• the colored part of the eye	• **controls** how much light goes into the eye through the pupil
pupil	• the black center part of the eye surrounded by the iris	• light passes through the pupil to the lens
lens	• behind the pupil, about the size of a pea	• **focuses** by changing shape very fast • lets us see things both up close and far away
retina	• the very back of the eye, behind the lens	• the **image** we see is created on the retina
optic nerve	• links the retina to the brain	• carries the image from the retina to the brain as an **electrical signal**
eyebrow **eyelash** **eyelid**	• the ridges of short hair that grow above the eye • the tiny hairs that surround the eye on the eyelid • the thin covering of skin that closes over our eye when we blink	• these parts protect the eye from damage that can come from many things including dirt, dust, bright light and drying winds

Why are the eyelashes, eyebrows and eyelids important to good eyesight?

_____

_____

The Sense of Sight

1. Fill in each blank with a word from the list.

pupil	lens	eyebrows	retina	pea
eyelids	brain	optic	blinking	iris	eyelashes

The colored part of the eye is called the _____. In the center of the eye is a

      a

black part called the _____. Behind both of these is the _____. This part

                b   **c**

of the eye is very small. It is only the size of a _____. The _____ is located

                                                    d   **e**

at the back of the eye. It is linked to the _____ nerve. This nerve sends messages

                                                f

to our _____. Our eyes are protected in several ways. Our _____ and

        g   **h**

_____ are hairs that help keep dust and dirt out. We can also protect our eyes

  i

by closing and opening our _____ quickly. This is also called _____.

                              j   **k**

2. Circle 🅣 if the statement is TRUE or 🅕 if it is FALSE.

T F   **a)** The optic nerve sends signals to our brain so we can understand what we see.

T F   **b)** The lens focuses by changing shape.

T F   **c)** Seeing different things can be fun but it isn't all that useful.

T F   **d)** The pupil controls how much light is let into the eye.

T F   **e)** The retina is where the image we see is created.

T F   **f)** Eyebrows have no purpose.

The Sense of Sight

3. How are the eyelashes and eyebrows similar and different?

_____

_____

4. Why does the **lens** of the eye need to be able to change shape?

_____

_____

5. Describe the **path** that light follows as it enters the eye. (Hint: this path ends when light hits the retina.) Tell what happens when light hits the retina. Look at the chart in the reading passage to help you.

_____

_____

6. Look at the diagram on the next page. Label the **parts of the eye.** Use the information from the reading passage, the Internet, or a book from the library to help you.

Extension & Application

7. **Helen Keller** was a very famous writer from the past who was also blind and deaf. Research to find out about Helen's life. Imagine that you had a chance to meet her. Using what you have learned from your research, write a **short story** describing your meeting.

> Think about the following: • **What kind of person was she?** • **How would you communicate with each other?** • **Are there any special questions you would want to ask?** • **How would she answer your questions?**

8. Many people wear glasses to improve their eyesight. **Interview two people** you know who wear glasses. They could be friends, family members or even yourself! Find out when they first got glasses, why they need them, and how they feel about them.

The Parts of the Eye

lens   iris   retina   pupil   optic nerve

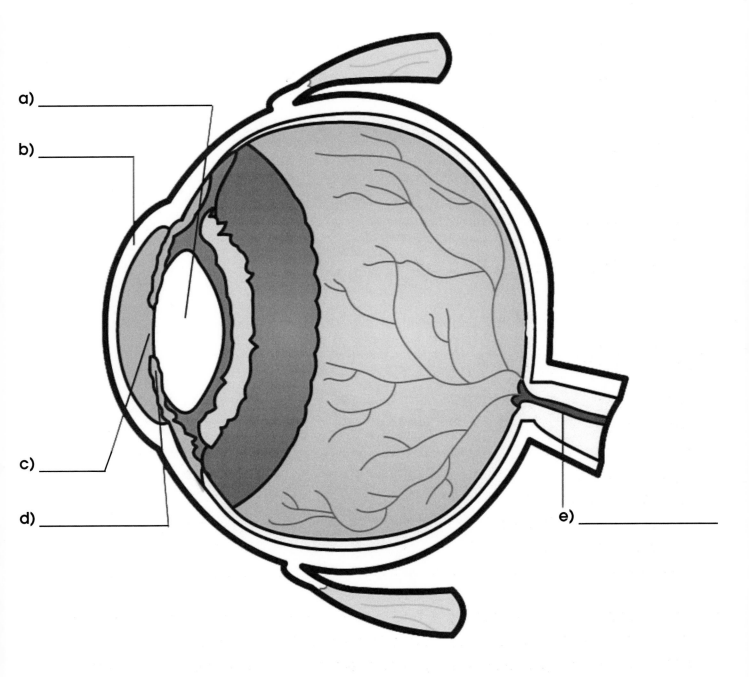

a) _____

b) _____

c) _____

d) _____

e) _____

The Sense of Hearing

1. You be the teacher! Someone has matched the words on the left to the definitions on the right. Are they correct? If **yes,** mark them correct with a check mark in the circle beside each. If **no**, write an X in the circle and correct the work by drawing an arrow to the correct definition. You may use a dictionary to help.

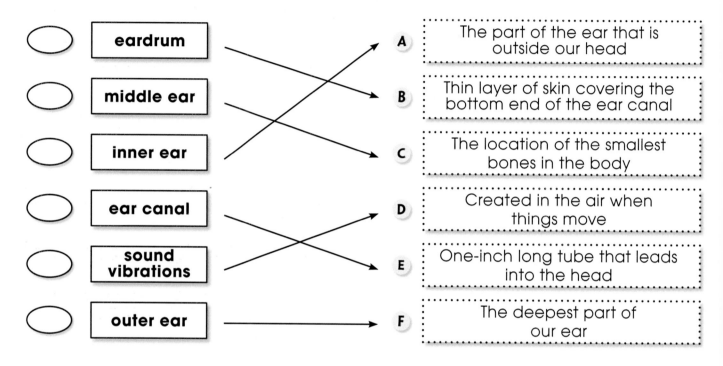

◯	eardrum	**A** The part of the ear that is outside our head
◯	middle ear	**B** Thin layer of skin covering the bottom end of the ear canal
◯	inner ear	**C** The location of the smallest bones in the body
◯	ear canal	**D** Created in the air when things move
◯	sound vibrations	**E** One-inch long tube that leads into the head
◯	outer ear	**F** The deepest part of our ear

2. Label the main <u>parts of the ear</u>. Use the words in the list.

inner ear   outer ear   middle ear   ear canal

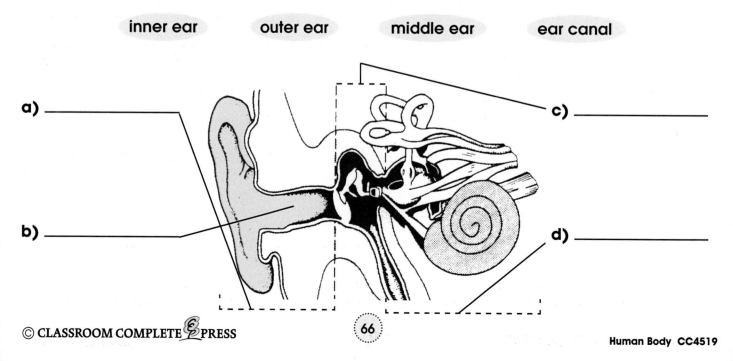

a) _____

b) _____

c) _____

d) _____

The Sense of Hearing

We hear sound when movement – like a door slamming – stirs the air around us and makes **vibrations** in the air. These vibrations move into our ears. They travel inside our head and are turned into **electrical signals**. These signals are sent to our brain, and a message telling us what the sound means is made. Sound waves travel very fast. This means we understand what we hear as soon as we hear it.

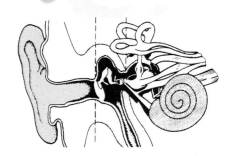

The parts of the ear are divided into three main areas: the outer ear, the middle ear, and the inner ear.

Ear Part	Description/Location	What It Does (Function)
outer ear	• the part of the ear that we can see on the side of the head • shaped like a big funnel • also called the **auricle**	• job is to **move as much sound as possible** into our head • the "funnel" traps sound so that it goes into the ear
ear canal	• a tube about one inch long that leads into the skull	• the sound travels down the ear canal to the eardrum
eardrum	• at the bottom of the ear canal • the beginning of **the middle ear**	• job is to **increase the power of the sound,** and move it deep into our head • vibrates like the top of a drum when sound hits it
stirrup, hammer and anvil	• all these parts are behind the eardrum • three of the smallest bones in the body, each named for the shape of the bone	• these bones vibrate when sound passes by
inner ear	• is deep inside our skull	• job is to **send the sound vibrations to the brain**
cochlea	• part of the inner ear • looks like a snail shell	• the sound travels into the cochlea and over the nerve cells
nerve cells	• located inside the cochlea • there are many of these tiny cells	• sound is turned into electrical signals • these signals are sent along the nerve cells to the brain

Why are sound vibrations important for hearing?

_____

_____

The Sense of Hearing

1. Use the terms in the list to answer each question. *Four* terms will be left over.

very slow	very fast	cochlea	ear canal
electrical signals	inner ear	middle ear	optic tube
optic nerve			

_____ **a)** Where is the eardrum found?

_____ **b)** Deep in the ear, what are sound vibrations turned into?

_____ **c)** Sending signals to the brain is the job of which part of the ear?

_____ **d)** How fast or slow does sound travel?

_____ **e)** What is the name of the tube that leads into our ear?

2. Fill in each blank with a term from the list. *Two* terms will be left over.

ear canal	eardrum	three	auricle	skull	anvil
inner ear	hammer	cochlea	stirrup	seven	middle ear

The outer ear includes the _____ which we can see on the outside of the head.
                                    a

The _____ is a one inch long tube that leads into the ear. The _____
        b   **c**

is at the bottom of the ear canal. This is where the _____ begins. Behind
                                                            d

the eardrum are _____ very small bones. They are called the _____,
                        e   **f**

_____ and _____. The _____ is deep in our skull.
        g   **h**   **i**

It contains the _____ and nerve cells.
                        j

NAME: _____

The Sense of Hearing

3. Why is the **eardrum** important?

_____

_____

4. What is the job of the **inner ear**?

_____

_____

5. How does sound travel deep into the ear? Use the terms **inner ear, outer ear, middle ear,** and **ear canal** in your answer.

_____

_____

Extension & Application

6. Research **two** different **musical instruments.** You could choose the piano, guitar, tuba, drum, flute or any instrument that you like. Just make sure that the instruments you choose are very different from each other. Find out what materials each is made of. What kinds of sounds do they make? How are they made? Be sure to include a picture or illustration of each instrument. Write down these findings in the chart on the next page. Tell why you chose these instruments.

Bonus: Can you find out when each instrument was first **invented?** Who are some well-known musicians who play these instruments?

7. You have probably heard that **loud noises can damage your hearing.** Loud sounds can be a real problem if they last for a long time. Do some research to find out how high volume sounds can injure your hearing.

Here are some questions for you to think about: • **What part or parts of the ears get damaged by loud sounds?** • **How are they damaged?** • **Can ears be damaged by sound in a very short period of time (like a loud crash or bang), or does damage happen only over a longer period of time?**

You can look for information in the library or on the Internet. Write down your information in a one-page report.

NAME: _____

Musical Instruments and Their Sounds

	Instrument 1:	Instrument 2:
Why chosen?		
Materials		
Sounds		
Picture		
Bonus: Invented when?		
Bonus: Famous Musicians		

NAME: _____

The Respiratory System

1. Match the words on the left to the definitions on the right. You may use a dictionary to help.

flap	Damp or a little bit wet	**A**
epiglottis	A covering that moves	**B**
carbon dioxide	The flap of skin covering the windpipe	**C**
cough	The waste gas we breathe out (exhale)	**D**
moist	To force air out of our lungs quickly	**E**

2. The respiratory system has **two** main jobs. Can you guess what they are? Think about how your body works and what you already know about breathing. Then **underline** your two guesses from the list below.

clean the windpipe   **clean the lungs**   **swallow**
cough   **breathe in oxygen**   **sneeze**
breathe out carbon dioxide

3. In the chart below list some of the things you have been able to do today because of your respiratory system. Try to list **at least five** different things.

1. _____

2. _____

3. _____

4. _____

5. _____

Thanks, Respiratory System!

The Respiratory System

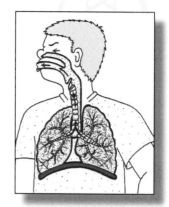

Humans need an important gas called **oxygen** to live. Without oxygen we would die in a few minutes. We get oxygen from the air around us. Our **respiratory system** brings the air into our body where we can use it. At the same time, our body needs to remove a waste gas called **carbon dioxide**. Our respiratory system does this job too. We breathe in oxygen and breathe out carbon dioxide 24 hours a day, all our lives, without having to think about it.

Mouth, Nose and Nasal Cavity

Air comes into the respiratory system through our mouth and nose. Dirt and **germs** are trapped by small hairs and **mucus** in the nose and nasal cavity. The mucus also **moistens** the dry air. The air is heated by passing over warm blood **vessels.** Did you know that the air we breathe in through our **mouth** does not get as **clean**, **moist** or **warm** as air entering through our nose? This is why it is healthier to always breathe through your nose.

STOP

What are the three reasons it is healthier to breathe through your nose?

_____

_____

Trachea and Epiglottis

You may have heard of the word "windpipe". The windpipe is also called the **trachea**. It is the tube that connects the upper respiratory system to the lungs. The trachea is made of rings of hard **cartilage tissue.** If you touch the front of your neck you can feel the trachea under your skin. It is very important that food or liquid never gets into your lungs. To keep this from happening, your windpipe is covered by a **flap** of skin called the **epiglottis**. The epiglottis **closes** when we eat or drink. If something enters accidentally, we **cough** to get it out.

The Lungs

The lungs are the most important part of our respiratory system. We learn about the lungs in the next section.

The Respiratory System

1. **Circle** **T** if the statement is TRUE or **F** if it is FALSE.

T F **a)** Carbon dioxide is the gas we need to live.

T F **b)** We breathe without having to think about it.

T F **c)** The windpipe is also called the trachea.

T F **d)** The windpipe is made of rings of strong muscle.

T F **e)** Without oxygen we would die in a few minutes.

T F **f)** Air is warmed, moistened and cleaned in our nasal passage.

T F **g)** Mucus in our nose moistens the air we breathe.

T F **h)** The epiglottis *closes* when we breathe and *opens* when we eat or drink.

2. **Fill in each blank with a term from the list. One term will be used <u>twice</u>.**

cartilage	germs	nose	lungs	trachea
moistened	flap	epiglottis	blood vessels	

Air comes into our _____ through our nose and mouth. It is healthier to breathe
a

through your _____. This is because air is _____ by nasal mucus.
b c

Mucus also traps dirt and _____. Air is heated by passing over _____. Air
d e

moves into the body down the _____. This is also called the windpipe. The
f

windpipe is made of strong _____. The top of the windpipe is covered by a
g

_____ of skin called the _____. This has the important job of
h i

keeping food and liquids out of the _____.
j

The Respiratory System

3. What does our **respiratory system** do for us?

_____

_____

4. Why is the **epiglottis** important?

_____

_____

5. What are the **two** main jobs of nasal mucus?

_____

_____

Extension & Application

6. Below is a list of terms from the reading. Choose **five** words that link together well. Then write a **short story** about the trip air takes as it moves towards the lungs. Pretend you are the air as you write your story. What is your trip like? Where do you begin your trip? Where does it end? What do your surroundings look like?

nasal cavity	mucus	warm	flap
epiglottis	tongue	nose	blood vessels
surprised	windpipe	slippery	scratchy
help	travel	moistens	drink
		cough	

7. When we **exercise,** we breathe harder and faster than when we are not active. Find out why this happens by researching on the Internet or in an encyclopedia.

The Respiratory System-Lungs

1. Use your dictionary to look up the meanings of the words below. Write the definitions in the space beside each word.

inhale	
exhale	
sac	
branches	
alveoli	

2. **Use the words from Question 1 above to complete the statements.**

   a) When we breathe in, we _____. When we breathe out, we

   _____.

   b) The _____ are small _____s that fill with air when we

   inhale.

   c) Our lungs are like trees. This is because one main "trunk" _____ into

   many smaller "twigs".

NAME: _____

The Respiratory System – Lungs

A human has a pair of lungs. Our lungs are the most complicated part of the respiratory system. Our lungs fill our chest and are protected by our ribs. Did you know that each lung is about the size of a football? When a lung fills with air it gets much larger. To feel your lungs get bigger, take a deep breath. Feel how your chest changes shape and gets bigger when you breathe air in (**inhale**). Your lungs **expand** as they fill with air.

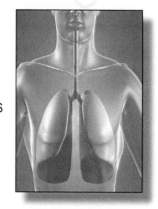

Our Lungs Look Like a Tree

A good way to picture our lungs is to think of a tree. Lungs have a solid trunk, just like a tree. Lungs also have **branches** that get smaller the further up you look. Our lungs are like an *up-side-down* tree. The **trachea (windpipe)** is like the trunk. It separates into two main branches, called the **bronchial tubes**. Each branch, or tube, leads into one of our lungs. These tubes branch many more times into thousands of "twigs". At the end of each "twig" is a small **sac** called the **alveoli.** The alveoli are like balloons; they fill with air when we breathe in. The inside **surface** of the alveoli is covered in tiny blood vessels. The blood vessels move **carbon dioxide** from the bloodstream into the lungs. Then, this gas moves out of the lungs as we breathe out (**exhale**). The blood vessels also take **oxygen** from the lungs to every part of the body.

Our Lungs Work Like a Train Station

The job of our lungs is to take gases in and out of our body. We can also say that the lungs **transfer** gases. Imagine your lungs working like a busy train station. With every breath, two kinds of *passengers* (oxygen and carbon dioxide) travel in and out to their *destination*. We breathe oxygen into our lungs. Then it travels from our lungs into our **bloodstream**. Carbon dioxide travels from our bloodstream into the lungs. Then we breathe it out.

STOP

What is the main job of the lungs?

_____

_____

NAME: _____

The Respiratory System - Lungs

1. Fill in each blank with a term from the list. **Two** terms will be left over.

air	chest	oxygen	transfer	pair	carbon dioxide
lungs	branches	sac	ribs	expand	exhale

The job of our lungs is to _____ gases. Humans have a _____ of
        a   **b**

lungs. Our _____ help protect our lungs from damage. Our lungs get bigger,
            c

or _____, when they fill with _____. The windpipe _____
   d   **e**   **f**

into two bronchial tubes. These tubes branch into many smaller "twigs". At the end of

each "twig" is a _____ that is like a balloon. This is where _____ from
                 g   **h**

the air moves into our bloodstream. It is also where _____ is moved from the
                                                     i

bloodstream into the lungs. Then, with our _____ we breathe the carbon
                                           j

dioxide out of our body.

2. (Circle) **T** if the statement is **TRUE** or **F** if it is **FALSE**.

T F **a)** Our lungs are shaped like soccer balls.

T F **b)** As we inhale, air moves out of our lungs.

T F **c)** The main job of the respiratory system is to transfer two gases in the body.

T F **d)** Carbon dioxide is taken out of our body when we breathe.

T F **e)** The insides of the alveoli are covered in tiny blood vessels.

T F **f)** From our lungs, oxygen is carried to all parts of our body.

T F **g)** Oxygen is a gas that is in air.

The Respiratory System - Lungs

3. In what ways do the lungs look like trees?

_____

_____

4. Why do we say that the lungs work like a train station?

_____

_____

5. Is the **bloodstream** important for the respiratory system to do its job? Why or why not?

_____

_____

6. Label the respiratory system using the terms in the list. Use information from the reading, the Internet, or a book from the library to help you.

windpipe   mouth   nose   bronchial tubes   lungs

a) _____

b) _____

c) _____

d) _____

e) _____

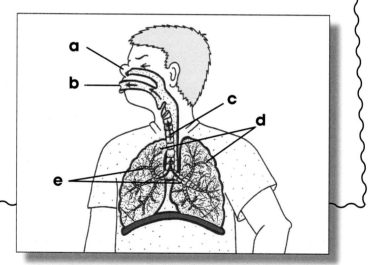

Extension & Application

7. The **diaphragm** is a large muscle just below the lungs. It is important for breathing. Do some research to find out how this muscle helps us breathe.

8. Read about the **Heimlich Maneuver.** What is it? Find out how it works with the respiratory system to save people who are choking.

Organ System Poster

We have learned about two important systems in the human body – the NERVOUS SYSTEM and the RESPIRATORY SYSTEM. You may remember that the nervous system is a lot like a computer system. The lungs (in the respiratory system) work like a train station.

For this activity, you will **create a poster** with important facts about one of these systems.

YOUR POSTER SHOULD INCLUDE:

- a catchy title to get the reader's attention
- an illustration that shows what the system looks like (be sure to label all the parts!)
- the main job or jobs of the system
- an explanation of:

    why the nervous system is like a computer system OR **why the lungs work like a train station**

- any other interesting facts that you find

Begin by collecting important **facts** about your system. You may use the reading passages, the Internet, or other resource materials to find your information. Make your poster on a piece of Bristol board. Make it colorful, neat and organized.

When you have it finished, **share** your poster with the class.

How Much Air?

We have learned that when you INHALE, your lungs fill with air, and that when you EXHALE, air is pushed out of your lungs. This activity will show you how much air your lungs can hold.

Work with a partner.

FOR THIS ACTIVITY, you will need: • a shallow tub • a 2-gallon pop bottle, filled with water • a flexible tube • a marker

STEPS:

1. Set up your materials as follows. (Your teacher may do this for you.)

   a. Make sure the bottle is filled up with water so there is no air in it.

   b. Put one end of the tube in the bottle of water.

   c. Partner A holds the bottle right-side up over the tub. (No water should fall out if you hold it up straight.)

   d. Partner B holds onto the other end of the tube.

2. Partner B takes a deep breath in, and then blows into the tube. Keep blowing into the tube until you have run out of air. (As you do this, you will see water coming out of the bottle. The air pushes the water out.)

3. The air that you will see inside the bottle is the air that was in your lungs. We call this your **lung capacity.** With your marker, put a small line on the bottle where the water begins.

4. Switch roles and repeat Steps 1 to 3.

5. Record your results on a graph like the one below.

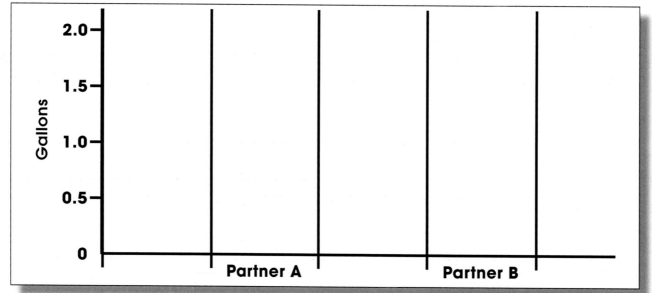

Taste and Smell

How connected are our senses of taste and smell?

We have learned that without our sense of smell, our sense of taste is weak. In other words, food tastes stronger when we can smell it, too.

Now it's time to find this out for yourself! Work with a partner.

FOR THIS ACTIVITY, you will need: • **2 pieces of raw potato (on a plate or paper towel)** • **2 pieces of raw apple** • **a blindfold**

STEPS:

1. Wash your hands before you begin.

2. Be sure that the pieces of potato and apple are all about the **same size**. **One** piece of potato and **one** piece of apple should be on a desk or table in front of you.

3. Partner A puts on the blindfold. (Partner B stands close by to help Partner A.)

4. Partner B moves the pieces of apple and potato on the desk so that Partner A does not know which is which.

5. Partner A holds his or her nose **shut**. Then they pick up one of the pieces of food and taste it. What are they tasting – the potato or the apple?

6. Partner A repeats Step 5 with the second piece of food.

7. Partner B tells Partner A whether they were correct.

8. Switch roles and repeat Steps 1 to 7.

Record your results below. What did the foods taste like? Could you tell the difference between them?

_____

_____

Did your experience **support** what you have learned (that taste and smell are connected)?

_____

_____

Memory Games – Touch and Sight

Part 1 - TOUCH

YOU WILL NEED:

- A small box or bag that you cannot see through. It should have an opening large enough for a hand.
- 12 different objects with a variety of textures (i.e., soft, scratchy, smooth, lumpy, etc.) and sizes
- a blindfold

Your teacher will give you these things. All the objects should be in the box or bag.
You must *not* know what the objects are.

HOW TO PLAY:

1. Take 30 seconds to feel the objects in the box/bag. Try to guess what they are with only your sense of touch.
2. In your notebook, list what you think the things in the bag are.
3. Look in the box/bag. Beside your wrong guesses, write the correct answers so that your list is complete and correct. Are you surprised by your results?

Part 2 - SIGHT

YOU WILL NEED:

- 15 to 25 different objects or pictures
- A table or small tray to put the objects on
- A cloth to cover the objects

Your teacher will give you these things.
You must *not* know what the objects are before you begin.

HOW TO PLAY:

1. Take the cloth off the collection of objects. Look at the collection of objects for one minute.
2. Cover the objects with the cloth.
3. In your notebook, list what you remember seeing.
4. Look at the objects again. Make any changes to your list so that it is complete and correct. Are you surprised by your results? Did you remember everything?

Crossword Puzzle!

Across

2. The lungs work like a _____
6. The two kinds of nerves are sensory nerves and _____ nerves
8. The colored part of the eye
9. Helps us see near and far
10. Connects the brain to the spinal cord
11. The ear _____ is the tube that leads into the ear
12. Feeling pain is important for our _____
14. A single bone in the spine
15. Our mouth and nasal cavity are _____ inside our head
18. The _____ nerve links the retina to the brain
19. Nerve _____ are also called neurons
20. The spinal cord is a thick _____ of nerves

Down

1. The cerbellum is important for balance and _____
3. The back of the eye where the image is made
4. The receptors in our _____ allow us to smell
5. Sound enters the head through the _____ ear
7. Nerves carry _____ to and from the brain
13. The spinal cord is _____ by the vertebrae
16. We breathe in the gas called _____
17. The nervous system is like a _____

Word List

bundle	outer
train station	posture
lens	nasal cavity
vertebra	messages
cells	computer
motor	protected
iris	retina
brain stem	oxygen
canal	connected
survival	optic

Human Body CC4519

Word Search

Find all of the words in the Word Search. Words are written horizontally, vertically, or diagonally, and some are even written backwards.

nervous system	lens	motor nerves	vibration
retina	cerebellum	brain	ear canal
electrical signal	windpipe	cochlea	blink
respiratory system	optic nerve	iris	auricle
pupil	vertebrae	spinal cord	pressure
cerebrum	neuron	sensory receptor	pain
sight	eardrum	focus	sensation

```
S P A S D C E R E B R U M Q N W E R
T I U E R U S S E R P V B J E I H F
U Y G P I O Z X C A U R I C L E A D
B V C H I X U Y T I R E W Q E A S P
M C K K T L L Z X N C V B N C M A S
R E S P I R A T O R Y S Y S T E M Q
W R E V E R T E B R A E R T R Y U I
Z E A R C A N A L X C V B N I M P N
X B C V B N M Q W E R T Y E C U I O
Z E X D C C O C H L E A V A A V N I
Q L W R E R T Y U I O R M R L P O T
A L S O N E U R O N E E D D S F I A
Z U X C C V B N M N T B C R I V T R
Q M W L E R T Y C S Y L A U G Z A B
Z W E A X C V I Y B S I Q M N W S I
X C V N B D T S F G H N Z X A C N V
C Q W I E P S D Z X C K V Z L X E S
L Q W P O U E X C D F C V B N M S R
S E N S O R Y R E C E P T O R Q E F
A I N V S D F S E V R E N R O T O M
X C R S V B N M Q W E G H J I C A S
J E H I W I N D P I P E G N U F D S
N X C C S V F G H J K L A S Q W E C
```

After You Read 📖

Comprehension Quiz

32

Part A

1. (Circle) **T** if the statement is **TRUE** or **F** if it is **FALSE**.

8

T F **1)** Our vertebrae protect our spinal cord from damage.

T F **2)** Nerve cells carry messages to and from the brain so that we can understand the world around us.

T F **3)** The three parts of the brain are called the cerebrum, cerebellum and brain stalk.

T F **4)** The iris is the colored part of the eye.

T F **5)** The sense of taste is *not* connected to the sense of smell.

T F **6)** Our lungs expand when we exhale.

T F **7)** The epiglottis covers the top part of the ear canal.

T F **8)** The small bones in the inner ear vibrate when sound passes by.

Part B ....

Label the nervous system in the diagram below. Use the words in the list.

spinal cord   nerves   brain

6

a) _____

b) _____

c) _____

SUBTOTAL: /14

After You Read

Comprehension Quiz

Answer each question in complete sentences.

1. Name the **three** main parts of the **nervous system**. Describe what each part does.

   (6)

   _____

   _____

   _____

2. Name **one** part of the **brain**. Describe where it is in the brain, and one main job it has.

   (3)

   _____

   _____

   _____

3. Name **one** of the two kinds of **nerves** in the human body. Describe what they do in the body.

   (2)

   _____

   _____

   _____

4. Describe what happens with **oxygen** and **carbon dioxide** in the respiratory system. Use the terms **inhale, exhale, lungs** and **bloodstream** in your answer.

   (4)

   _____

   _____

   _____

5. Describe how our body feels **pain**. Use the terms **skin, sensory receptors, message** and **brain** in your answer. Why is it important that we can feel pain?

   (3)

   _____

   _____

   _____

SUBTOTAL: /18

The Circulatory System – Blood Vessels

1. Complete each sentence with a word from the list. Use a dictionary to help you.

aorta	veins	circulates	capillary	pressure

a) The smallest blood vessel in the body is called a _____.

b) Blood pumping through our arteries causes _____ on them.

c) _____ are important because they carry blood from the body back to the heart.

d) Our blood _____ through our body in our blood vessels.

e) The _____ is the biggest artery in the body and is attached to the heart.

2. On the left side, list what you already know about the circulatory system. (Hint: it includes the heart and blood.) Then, list some questions you might have. Try to list at least three things in each column.

What I Know About the Circulatory System	Questions I Have About the Circulatory System

📖 *Reading Passage*

The Circulatory System – Blood Vessels

The circulatory system has two main jobs: **1)** carrying oxygen and food to all the cells of the body, and **2)** carrying waste away from the cells to be removed from the body. This system gets its name because blood moves, or **circulates**, to every part of the body. The three main parts of the circulatory system are the **heart, blood** and **blood vessels**. Let's start by learning about blood vessels.

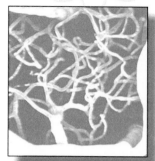

STOP

What are the three main parts of the circulatory system?

_____

_____

There are three different kinds of blood vessels in the human body. All blood vessels are tubes that carry blood. These are arteries, veins and capillaries.

Type of Blood Vessel	What They Do
Arteries are the largest blood vessels. They have very tough, flexible and thick walls. The biggest artery in the body is called the **aorta**. It is the size of a thumb and is connected to the top of the heart.	Arteries have the job of carrying blood away from the heart. (You can remember this because artery and away both begin with "a".) One reason their walls are thick is because there is a lot of **pressure** on them as blood is pumped from the heart. (Pressure is the force of pushing.)
Veins are smaller than arteries. They also have thinner walls.	The job of the veins is to carry blood from the body back to the heart. The walls of veins are thinner because there is less pressure on the blood going back to the heart.
Capillaries are the smallest of all the blood vessels. Their walls are only one cell thick.	Capillaries are all over our body. They go out to every cell. Their very thin walls make dropping off **oxygen** and picking up **wastes** easy.

NAME: _____

The Circulatory System – Blood Vessels

1. Fill in each blank with a term from the list.

arteries	capillaries	waste	circulatory	circulates
tubes	heart	veins	blood	food

The _____ **a** system carries oxygen and _____ **b** to our cells, and

takes _____ **c** away. This system gets its name because our blood _____ **d**

(moves) around our body. There are three main parts to the circulatory system. These

are the _____ **e**, blood and blood vessels. Blood vessels are _____ **f** that

carry _____ **g** around the body. From largest to smallest our blood vessels are

_____ **h**, _____ **i** and _____ **j**.

2. a) <u>Underline</u> the words and ideas that describe what the circulatory system does.

circulates   helps us think

cleans out waste   brings oxygen to the cells

pumps blood   helps us hear sound

carries waste to our lungs   helps keep us healthy

b) (Circle) the words that are parts of the circulatory system.

bones	circles	arteries	marrow
blood	heart	pupils	vessels
lungs	aorta	capillaries	veins

The Circulatory System – Blood Vessels

3. Why do the walls of **arteries** need to be thick and elastic?

_____

_____

4. Why are the walls of **capillaries** only one cell thick?

_____

_____

5. Does the statement describe arteries, veins or capillaries? Put an X in the correct box.

Feature	Arteries	Veins	Capillaries
a) The smallest blood vessels			
b) The aorta is one of these			
c) These carry blood to the heart from the body			
d) These carry blood away from the heart to the body			
e) These go to the cells and drop off oxygen			
f) The walls of these can be less thick because they are under less pressure than arteries			

Extension & Application

6. What is your **favorite animal**? Research to find out about its circulatory system. Look for information in library books or on the Internet. Then, **compare** the circulatory systems of your animal and a human. What is the same? What is different? Record your findings in a Venn Diagram, like this one:

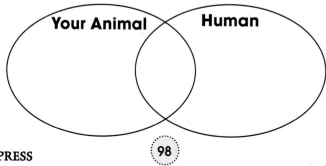

NAME: _____

The Circulatory System - Heart

1. Complete each sentence with a word from the list. Use a dictionary to help you.

pump   oxygenated   involuntary   cardiac   chamber

a) Blood that is _____ has a lot of oxygen in it.

b) A _____ is a separate space, like a room. The heart has four.

c) The heart muscle works without us thinking about it so it works _____.

d) _____ is another word for the heart and things related to it.

e) The heart is like a _____ because it pushes blood around the body.

2. Which term completes the sentence? Circle your answer. You may use your dictionary to help you.

a) Our heart _____.

beets   beats

b) The heart is about the size of a _____.

fist   first

c) Heart muscle is also called _____ muscle.

cardiac   carp

d) The heart is divided into sections called _____.

rooms   chambers

e) Muscle that works without us thinking about it is called _____.

voluntary   involuntary

The Circulatory System – Heart

Y our heart is only the size of your **fist**, but it is one of the strongest muscles in your body. That's a good thing because it has a very important job. The heart **pumps** blood through your whole body every moment you are alive.

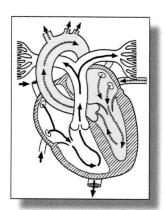

Facts About the Hardworking Heart
- Most hearts beat 75 to 80 times every minute.
- Your heart beats over 100,000 times every day.
- Blood travels around the body 1,000 times a day.
- If you live into your 70's, your heart will have beat over 3 billion times.

The heart is made of **involuntary muscle** tissue called **cardiac** muscle. Involuntary muscle works without us telling it to. This means that you do not have to tell your heart to beat. It does it without you even noticing. This is important because it would be impossible to remember to make your heart beat thousands of times an hour.

The Heart Is a Pump

The heart is really two pumps in one. The two parts work together as a team to send blood all around the body. The heart is divided into two sides. Each of these sides is divided into two chambers. This means the heart has four main areas.

The left side pumps blood that has come from the lungs and is filled with oxygen. Blood filled with oxygen is bright red. This **oxygen-rich (oxygenated)** blood is going to the body to feed every cell. The right side of the heart pumps blood back to the lungs where it can get more oxygen. This blood is called **oxygen-poor (deoxygenated)** and is a very dark red color.

STOP

What is the main job of the heart?

_____

_____

The Circulatory System – Heart

1. **Put a check mark (✓) next to the answer that is most correct.**

   a) **Why does the heart have to beat throughout our whole life?**

   ○ **A** It would get bored with nothing to do if it didn't.

   ○ **B** If our blood does not deliver oxygen to our cells they will die.

   ○ **C** Our heart signals our brain when to breathe.

   ○ **D** It doesn't beat all our life. Once we are older than 70 the heart doesn't beat.

   b) **Why is it important that cardiac muscle is involuntary?**

   ○ **A** Involuntary muscle works faster.

   ○ **B** Involuntary muscle is stronger.

   ○ **C** Involuntary muscle works without us telling it to.

   ○ **D** Involuntary muscle does not need oxygen to live.

2. **Fill in each blank with a word from the list. Two words will be left over.**

blood	two	four	deoxygenated	bright
lungs	cells	dark	oxygenated	poor

The heart works not one, but _____ pumps. The left side pumps blood to
                                    a

the _____. This blood is _____ because it is filled with oxygen. The
      b   **c**

oxygen makes this blood _____ red. The right side pumps blood back to the
                          d

_____ to drop off waste and pick up oxygen. This blood is _____ red.
      e   **f**

It has much less oxygen so it is called _____. It can also be called oxygen -
                                          g

_____.
      h

The Circulatory System - Heart

3. Why is it important that the heart beats without us telling it to beat?

_____

_____

4. What is the difference between bright red blood and dark red blood?

_____

_____

5. Here are some new words from the reading. Write each word beside its meaning. Use the information in the reading passage to help you.

   chamber   oxygenated   deoxygenated   involuntary   pump

_____ **a)** The heart is one because it pushes blood around the body.

_____ **b)** A separate place (the heart has four of these)

_____ **c)** Filled with oxygen

_____ **d)** Something that works without us having to think about it first

_____ **e)** Something with very little oxygen in it

6. Look at the diagram on the next page. It is a **heart**. Follow the instructions to show how blood flows through the two chambers of the heart.

Extension & Application

7. What do you know about **heart attacks?** They are very dangerous. People can die from them. There are other **heart diseases** too. Research heart attacks or another heart disease. Look for information on the Internet and in books from the library. Record your information in the chart (on page 104). Here are some questions you should find the answers to:

   • **What happens to the heart?** • **What are the symptoms?** • **How is it treated?**
   • **Are there some people who are more likely to get it than others?**
   • **What can people do to avoid getting it? What can people do to keep their heart healthy?** • **Any other interesting facts**

NAME: _____

The Heart

Look at the diagram of the heart. You can see how blood flows through the two chambers of the heart. You can tell which chamber is which by looking at the **arrows**. The arrows show which way the blood is circulating. Some blood is going into the heart. Some blood is going away from the heart.

REMEMBER:

Arteries take blood **away** from the heart.

Veins take blood **into** the heart.

a) Color the blood moving through the left chamber **red**.

b) Color the blood moving through the right chamber **blue**.

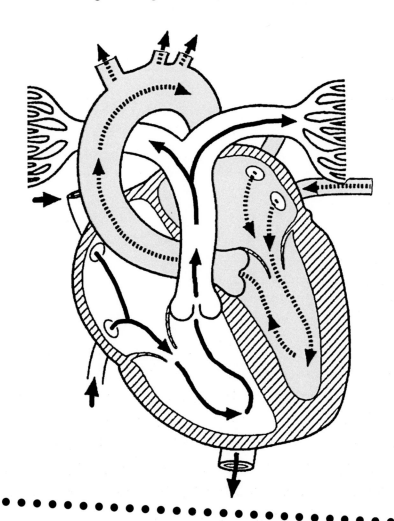

After You Read

A Disease or Other Illness of the Heart

NAME OF ILLNESS:	
1. What happens to the heart?	
2. What are the symptoms?	
3. How is it treated?	
4. Who is most likely to get it?	
5. How to keep heart healthy	
6. Other interesting facts	

The Circulatory System - Blood

1. **Match the word on the left to the definition on the right. You may use a dictionary to help.**

plasma	**A**	The part of our blood that makes it clot
platelet	**B**	To be protected from disease or harm
clot	**C**	The biggest part of blood that is made mostly of water
volume	**D**	A way that liquid is measured based on the amount of space it takes up
immunity	**E**	When blood thickens it does this

2. Write about a time you got hurt and **started to bleed.** Maybe you cut your finger. Maybe you fell and scraped your knee, or got a nosebleed. It could have happened recently or a long time ago. **Tell what happened and how you felt**. How much bleeding was there? What did you do to stop the bleeding? How long did it take to stop? How long did it take to heal?

_____

_____

_____

_____

_____

_____

_____

The Circulatory System – Blood

Blood is the red liquid that flows through our **arteries**, **veins** and **capillaries.** Its job is to carry oxygen and nutrients to the cells and carry wastes away from the cells. Blood is created in **marrow** inside our bones. The average adult has about 10 pints of blood in their body. Blood has water in it, but is thicker and **saltier** than water.

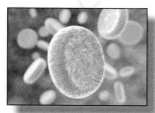

Parts of the Blood and What They Do

Blood may seem like just one thing, but it is really made of four main parts. These are red blood cells, white blood cells, plasma and platelets.

Blood Part	What It Does
Almost all of our blood cells are **red blood cells.** This is why our blood is red. These cells get their red color from the iron in them.	Red blood cells carry two important gases around our body – **oxygen** and **carbon dioxide.**
White blood cells make up only a small part of our total blood volume.	White blood cells are important for our **immunity.** They have the job of fighting diseases and all kinds of **germs.**
Plasma is the liquid that makes up most of our blood. Plasma is 95 percent water.	The main job of plasma is to hold the other parts of blood. Plasma makes blood **watery.**
There are fewer **platelets** in blood than white or red blood cells. Platelets are still very important.	Platelets thicken and harden our blood in air. This is called **clotting.** It is our body's way of making sure we stop bleeding, covering our wounds to help them heal. Hardened clots become **scabs.**

STOP

Name the four main parts of blood.

_____

_____

NAME: _____

The Circulatory System - Blood

1. Fill in each blank with a term from the list.

red	plasma	four	iron	protecting	platelets
water	clots	white	immunity	scab	

Blood has _____ important parts. _____ is the liquid that makes up
            a   **b**

most of our blood. Plasma is made of mostly _____. _____ blood
                                                  c   **d**

cells are the most common kind of blood cell. The _____ in these cells gives
                                                        e

blood its red color. _____ blood cells have the important job of _____
                          f   **g**

us from diseases and germs. This is called giving us _____. When we hurt
                                                          h

ourselves and start to bleed, _____ stop the bleeding by making
                                    i

_____. This means that the blood thickens and dries into a _____.
     j   **k**

2. (Circle) **T** if the statement is TRUE or **F** if it is FALSE.

T F a) Platelets are the part of blood that fight germs and disease.

T F b) When blood thickens it is called clotting.

T F c) We have fewer white blood cells than red blood cells.

T F d) The heart is a voluntary muscle.

T F e) It would be impossible to remember to make our heart beat.

T F f) Blood with not much oxygen in it is called oxygen-poor or
        deoxygenated.

T F g) The human heart is the size of a football.

The Circulatory System – Blood

3. Why is it important that blood can clot and scabs can form?

_____

_____

4. Where in the body are blood cells made?

_____

_____

5. What would happen if we had too few white blood cells or none at all?

_____

_____

6. Match the **blood part** on the left with its **job** on the right.

scab		A	Holds the other parts of blood
plasma		B	Protects a wound and stops bleeding
red blood cells		C	Carry oxygen to the cells

Extension & Application

7. **Many people donate blood.** Research to find out how people donate blood where you live. Where do they go? Who do they see? Who is in charge? What happens? How is blood kept until it is needed? Where is blood stored? You may wish to interview someone in your family, or a friend, who has donated blood. Find out why they did it.

8. **Leukemia** is a serious blood disease. It is a cancer of the blood. Research to find out as much as you can about leukemia. What are its symptoms? How is it treated? Write down your findings in your notebook.

The Digestive System – Mouth to Stomach

1. Use the dictionary to look up the meanings of the words below. Write the definitions in the space beside each word.

saliva	
secrete	
pouch	
acid	
churn	

2. On the left side, list what you already know about the digestive system. (Hint: it includes the stomach and other organs.) Then, list some questions you might have. Try to list at least three things in each column.

What I Know About the Digestive System	Questions I Have About the Digestive System

The Digestive System – Mouth to Stomach

The human, like all animals, needs fuel to stay alive. Humans get fuel by eating food. Think about the food you eat. Maybe your favorites include apples, macaroni and cheese, hamburgers or ice cold milk. How can these things feed our tiniest cells? Something must happen to our food to turn it into fuel. This process is called **digestion.** There are four main parts to our digestive system. These are the mouth, stomach, large intestines and small intestines.

How Do the Mouth and Stomach Work to Begin Digestion?

The **mouth** is where digestion begins. When we take a bite of food we chew it into small pieces. Our teeth bite, grind and rip the food. The muscles of the mouth and **tongue** move the food and mix it with the **saliva** in our mouth. Saliva is a watery liquid that our mouth **secretes** to help digestion. One of the most important jobs of saliva is to start to break **starchy** food (like pasta) down into simple sugars. For the best digestion, food needs to be chewed many times.

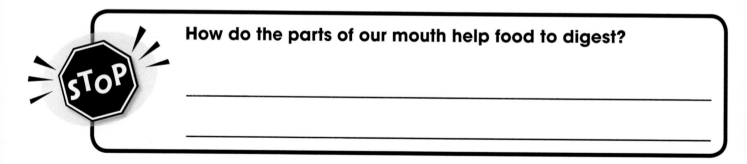

STOP

How do the parts of our mouth help food to digest?

_____

_____

Next, we swallow the chewed food and it continues on its path to the **stomach**. The food is pushed down our **esophagus** by **involuntary muscles**. Saliva helps here too because it makes food slide down more easily. The stomach is less than two feet below the mouth. The stomach is a pouch of muscle. When it is empty it is about the size of your fist.

Our stomach is very strong and works like a washing machine. The muscular lining churns and squeezes the food to break it down into a liquid. The stomach adds special **chemicals** to help digest the food. These chemicals are very **acidic** (full of acid). The acid helps break the food down fast. Food stays in the stomach for three to six hours. By the time it is ready to leave the stomach it has been turned into a thick liquid.

After You Read

The Digestive System – Mouth to Stomach

1. **Use the terms in the box to answer each question. Some terms will be left over.**

three	churns	four	muscle	football
saliva	nerves	fist	esophagus	

_____ **a)** What watery liquid helps break down starchy food?

_____ **b)** What is our stomach made of?

_____ **c)** What is the name of the tube that leads from our mouth to our stomach?

_____ **d)** How many main parts are there in the digestive system?

_____ **e)** How big is the stomach when it is empty?

_____ **f)** How does the stomach break down food?

2. **Fill in each blank with a word from the list. Some words will be left over.**

saliva	starch	few	four	teeth	sugar
tongue	many	salt	swallow	intestine	mouth

There are _____ main parts to the digestive system. These are the mouth,
                    a

the stomach, and the large and small _____s. Digestion begins in the
                                        b

_____. Our _____ rip and chew our food. Our _____
       c   **d**   **e**

moves the food around in our mouth. _____ is the watery liquid that moistens our
                                           f

food and makes it easier to _____. Saliva also helps break down _____
                                  g   **h**

into _____. It is healthiest when we chew our food _____ times.
           i   **j**

The Digestive System – Mouth to Stomach

3. What kind of muscles line the esophagus?

_____

_____

4. How does the stomach digest food?

_____

_____

_____

Extension & Application

5. Below is a list of words from the reading. Choose SIX words that link together well. Circle them. Then write a paragraph about a piece of pizza that is getting eaten by a hungry classmate. Write your story as if YOU are the piece of pizza! What happens? What would it be like if YOU were a piece of pizza getting eaten by a hungry classmate? Use the facts from the reading and your own ideas. Be creative!

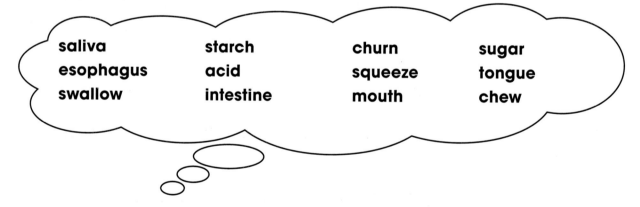

saliva	starch	churn	sugar
esophagus	acid	squeeze	tongue
swallow	intestine	mouth	chew

6. Most of us have been **sick to our stomach or vomited** at some time. Vomiting is unpleasant. However, it can also be helpful. It can actually protect us from great harm.

a) Research to find out how vomiting can **protect us from harm.** Look in books from the library or on the Internet.

b) Next, create a list of **ten poisonous things** around the house that could make us very sick. You will need to do some research to complete this activity.

The Digestive System – From Stomach to Fuel

1. **Use the dictionary to look up the meanings of the words below. Write the definitions in the space beside each word.**

diameter	
absorb	
enzymes	
coiled	
indigestible	
nutrients	
mineral	

2. **Use the words from Question 1 above to complete the statements. (Not all the words will be used.)**

   a) Special chemicals called _____ help break down our food.

   b) The large intestine gets its name from its _____.

   c) Things that are _____ are passed from our body as waste.

   d) The small intestine is very long but it fits inside us because it is _____.

The Digestive System – From Stomach to Fuel

A fter several hours in the stomach, food is only partly digested. Next, it moves into the **small intestine.** The small intestine gets its name because it is a tube that is small around (small in **diameter**). But the small intestine is also very long. It is close to twenty-five feet long! It fits in our body because it is very tightly **coiled** (like a curled up snake that is sleeping).

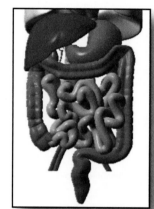

The Small Intestine

Chemicals, called **enzymes**, enter the small intestine to help digestion even more. These enzymes come from two important organs – the **liver** and the **pancreas.** Remember that the food was broken down by the stomach into very simple parts. These parts include sugars. These **nutrients** are absorbed into the walls of the small intestine. Then they move into the bloodstream. From the bloodstream, the **nutrients** are carried to all parts of the body. So you can see the important role of the small intestine. This is the place where most of our food's nutrients are made usable to our cells.

STOP

What is the main role of the small intestine in digestion?

_____

_____

The Large Intestine

From the small intestine, what is left of our meal passes to the **large intestine**. This organ gets its name because it is larger in diameter than its small neighbor. The large intestine is short – only about five feet long. The material left at this point is almost all **waste** that is **indigestible**. The main job of the large intestine is to take out important **minerals** and liquid from the waste. This moisture is absorbed back into the body through the wall of the large intestine. After several hours the waste is ready to leave the body.

NAME: _____

The Digestive System – From Stomach to Fuel

1. Put a check mark (☺) next to the answer that is most correct.

a) What two things are true about the small intestine?

- ○ **A** It is short and small in diameter.
- ○ **B** It is about twenty-five feet long and tightly coiled.
- ○ **C** It is about twenty-five feet long and very large around.
- ○ **D** It is the place where minerals and moisture are removed from waste.

b) Where do the enzymes that help digest our food come from?

- ○ **A** the liver and stomach
- ○ **B** the liver and large intestine
- ○ **C** the pancreas and stomach
- ○ **D** the liver and pancreas

c) What does it mean if something is indigestible?

- ○ **A** It takes a long time to digest.
- ○ **B** It is already digested.
- ○ **C** It cannot be digested.
- ○ **D** None of the above.

2. Fill in each blank with a word from the list. Some words will be left over.

liver	indigestible	nutrients	moisture	large
minerals	small	enzymes	cells	

The small intestine is where most of the _____ are taken from our food.
  _a

_____ are special chemicals from the pancreas and the _____.
  _b   _c

They help us digest our food. The material left over is mostly waste that is

_____. In the _____ intestine important _____ are
  _d   _e   _f

taken out of the waste. Some _____ is also absorbed back into the body.
  _g

NAME: _____

The Digestive System – From Stomach to Fuel

3. Why does the small intestine fit inside our body even though it is very long?

_____

_____

4. How long is the large intestine?

_____

_____

5. What happens to waste that our body cannot digest?

_____

_____

6. Match the word on the left with its job or definition on the right.

waste	A	Chemicals that help digestion
pancreas	B	The material that our body cannot use
enzymes	C	One of the organs that makes enzymes
water	D	Removed in the large intestine

Research & Extension

7. Read more about the **liver** or the **pancreas**. Find **at least five more** interesting facts about the organ you choose. Try to find out where it is in the human body. What does it look like? Draw a picture of it in the human body to show its size, shape and location. Include other interesting facts that you find.

The Excretory System – Skin, Liver and Lungs

1. **Use a dictionary to find the meanings of the words below. Write the definitions in the space beside each word.**

toxic	
poison	
pores	
exhale	
filter	

2. Choose **three** words from Question 1 above. Copy the format below into your notebook three times. For each of the three words you've chosen, sketch a picture to help you remember the meaning of the word. Then, write each word in a sentence that shows you understand the meaning of the word.

Word:

Picture Cue:

Own Sentence: _____

_____

_____

The Excretory System – Skin, Liver and Lungs

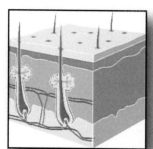

Our organ systems make many things we need to live. Did you know that they also create waste that is **toxic** to our body? The excretory system keeps our body clean of this **waste**. If it wasn't for this important system, this waste would poison us and we would die. The main organs of the excretory system are the **kidneys** and **large intestine**. We will learn more about them in the next section. Three other important organs of the excretory system are the skin, the lungs and the liver.

Body Organ	How It Helps to Keep Our Body Clean
Skin	The skin is the largest organ of the body. The skin helps get rid of waste in two ways: 1. Dead skin cells are shed from the body constantly. 2. **Moisture**, called **perspiration** or **sweat**, is removed through tiny holes in the skin. These holes are called **pores**. One of the main wastes in sweat is salt that the body does not need.
Lungs	**Carbon dioxide** is a waste gas. It is always being made by the body. In the lungs, it is taken out of the blood. Then it is taken out of the body every time we breathe out, or **exhale**.
Liver	The liver is a large organ on the right side of the body. It is in the upper **abdomen**. The average liver weighs five pounds and is reddish-brown in color. Its job is to **filter** or clean the blood. As blood flows into the liver, it cleans out salts, **acids** and old blood cells. It also gets rid of other **toxins**. Some examples are old medicines and other **chemicals** that get into our body from the environment.

Why is it important that our body can get rid of wastes?

_____

_____

The Excretory System – Skin, Liver and Lungs

1. **Circle T if the statement is TRUE or F if it is FALSE.**

   T   F   a) The skin is the largest organ of the body.

   T   F   b) The excretory system helps our body use its waste.

   T   F   c) The large intestine and kidneys are two important parts of the excretory system.

   T   F   d) The job of the liver is to filter and clean our blood.

   T   F   e) Perspiration and sweat are the same thing.

   T   F   f) The lungs are very important for breathing, but not for removing waste.

   T   F   g) The liver is one of the smallest organs in the body.

   T   F   h) If our body could not remove waste we would die.

   T   F   i) The liver cleans out salt and acid from the body.

2. **Fill in each blank with a term from the list. One term will be left over.**

salt	sugar	acid	pores	carbon dioxide
exhale	blood	perspires	dead	liver

The lungs remove _____ from our body. We put this back into the air when we
                        a

_____. The skin removes waste by shedding _____ skin cells. The skin
      b   **c**

also _____. This is also called sweating. Sweat comes out of the skin through
        d

_____. These are tiny holes in the skin. Sweat also has _____ in it. The
      e   **f**

_____ is reddish-brown in color. It is a large organ that cleans our _____.
      g   **h**

Salt, _____ and old blood cells are all removed here.
          i

The Excretory System – Skin, Liver and Lungs

3. **What is one of the main wastes in perspiration (sweat)?**

_____

_____

4. **Which organ cleans the body of chemicals from the environment?**

_____

_____

5. **Is the liver a large or small organ? How much does it weigh?**

_____

_____

Extension & Application

6. Below is a list of terms from the reading. Choose SIX words that link together well. Circle them. Then **write a paragraph** using all six words. Remember that a good paragraph has a topic sentence, main points and a concluding sentence. Use the Paragraph Organizer on the next page to help you.

salt	remove	holes	acid	pores	carbon dioxide
gas	clean	dead	toxic	exhale	lungs
blood	perspire	liver			

7. One of the jobs of our liver is to clean toxic chemicals out of the body. These chemicals come from the environment around us. There are many toxic chemicals that our liver protects us from. Do a bit of research to find out about some toxic chemicals. Make a list of **five toxic chemicals in your notebook**. Note where they come from and what their job is. Tell how they get into our air, water or food.

Paragraph Organizer

Topic Sentence	
Main Points **1.** **2.** **3.**	
Concluding Sentence	

📖 Before You Read

The Excretory System – Kidneys and Large Intestine

1. Match the word on the left to the definition on the right. You may use a dictionary to help you.

colon	**A**	Dampness or wetness
abdomen	**B**	The main liquid waste from our body
urine	**C**	A tiny piece or bit
particle	**D**	Another name for the large intestine
moisture	**E**	The middle part of our body

2. Which word completes the sentence? Circle your answer. You may need to use a dictionary.

a) Our excretory system can be called our _____ removal system.

waste   **food**

b) The kidneys are filled with millions of _____.

tubs   **tubes**

c) Our bladder is a _____ that holds urine.

porch   **pouch**

d) The kidneys send some of the cleaned water back to the _____.

sells   **cells**

e) The colon stores _____ waste from the body.

solid   **sold**

NAME: _____

 Reading Passage

The Excretory System – Kidneys and Large Intestine

The most important excretory organs in the body are the **kidneys** and the **large intestine**. We have two kidneys. They are small, dark red organs each about four inches long. If you have ever seen a kidney bean, you know the shape and color of your kidneys. Each kidney is located in the lower back on either side of our spine.

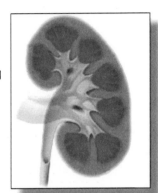

The large intestine is also called the **colon**. It is about five feet long. It is in the lower **abdomen**. Its shape is that of an up-side-down "**U**". Let's look at the chart below to learn more about these two important organs.

Body Organ	How It Helps to Keep Our Body Clean
Kidneys	The kidneys are the main waste removal organs. **Liquid** waste is taken out of the blood through millions of tubes in the kidneys. Some of the cleaned water is then returned to our cells. The rest of the water holds waste **particles**. This liquid waste is called **urine**. Urine drains from the kidneys into the **bladder**. The bladder is like a balloon, and can hold up to a pint of urine. Urine stays in the bladder until it is passed from the body.
Large Intestine	We have read that food moves from the stomach to the small intestine. Then it passes to the large intestine. In the large intestine, the food is called **solid waste**. Here, moisture is taken out of it. Finally, the waste is pushed out of the body through an opening at the end of the intestine.

STOP **What is the role of the bladder in removing liquid waste?**

_____

_____

The Excretory System – Kidneys and Large Intestine

1. **Circle** **T** if the statement is TRUE or **F** if it is FALSE.

T **F** **a)** "Colon" is another name for the large intestine.

T **F** **b)** Urine is the same as water.

T **F** **c)** Kidneys are important for circulation.

T **F** **d)** The kidneys are located on either side of our spine.

T **F** **e)** The kidneys are filled with millions of tiny tubes.

T **F** **f)** Three important excretory organs are the kidneys, the liver and the abdomen.

T **F** **g)** The colon is twenty-five feet long and is coiled in our body.

T **F** **h)** Urine drains from the kidneys into the bladder.

2. **Fill in each blank with a word from the list. One word will be left over.**

water	passed	liquid	balloon	cell
kidney	urine	particle	pint	bladder

Our two _____ s are important for removing waste. They remove _____
       a   **b**

waste. _____ goes through the millions of tubes in the kidneys. Here, some water
       c

is cleaned and sent back to the _____ s. What is left has waste _____
                                     d   **e**

s in it. This is called _____. Urine drains into the _____. This organ is
                          f   **g**

like a _____ that gets bigger as it fills up. Urine stays in the bladder until it is
       h

_____ from the body.
       i

The Excretory System – Kidneys and Large Intestine

3. What are the two most important organs of excretion?

_____

_____

4. How much urine can a bladder hold?

_____

_____

5. Which of the below are true statements about the excretory system? Use an X to mark your choices.

Excretory System	True
a) If our excretory system did not work we would die.	
b) The kidneys and the bladder work together to get rid of liquid waste.	
c) Liquid waste is called urine.	
d) The colon helps the kidneys remove urine.	
e) Salt is a body waste.	
f) Each of our kidneys weighs about five pounds.	

Extension & Application

6. Have you ever heard of the term **dialysis**? This is a medical treatment for the kidneys. People have to use this treatment if their kidneys don't work. Research to find out what dialysis is and how it works. Are there diseases that can damage the kidneys?

📖 Before You Read

The Reproductive System

1. **Use the dictionary to look up the meanings of the words below. Write the definitions in the space beside each word.**

species	
hormone	
fetus	
navel	
estrogen	
testosterone	

2. **Use the words from Question 1 above to complete the statements. Not all words will be used.**

a) The male reproductive hormone is called _____.

b) The female reproductive hormone is called _____.

c) Cats, humans and bald eagles are all examples of _____.

d) Some people call it a belly button but its proper name is _____.

e) A developing human still in its mothers's body is called a _____.

NAME: _____

The Reproductive System

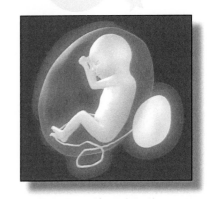

The human reproductive system is important for one good reason: it creates new people. Not every person has children, but some people must have children for our **species** to survive into the future. The way humans reproduce is called **sexual reproduction**.

The reproductive system has three main jobs:
1. To create **egg** and **sperm** cells
2. To grow the developing baby
3. To make chemicals called hormones

Why is human reproduction important?

_____

_____

The **male** body makes, or **produces**, sperm cells. The **female** body produces egg cells. These cells are smaller than a grain of sand. **Fertilization** happens when the egg and sperm cells join together in the female body. Then, cells start to divide and grow. Finally, these cells will become a baby.

An unborn baby (or **fetus**) grows inside its mother's body for nine months. It begins as just two cells. Then, these two cells divide into four cells. The four cells become eight cells. The cells keep growing and dividing. As time passes, the fetus looks more and more like a person. Every part of the human body is grown in this time. The baby grows arms and legs. All its inside organs are formed and grow. After three months of growth, the fetus is only about three inches long. It starts to move more and more. The fetus gets food and oxygen from the mother's body. These come through a tube, called the **umbilical cord**. This cord was attached to *you* at your belly button! Your belly button is also called your **navel**. After nine months the baby is born. A newborn baby is about 20 inches long. It weighs seven to nine pounds.

The reproductive system also makes **hormones**. Hormones are special chemicals that help our bodies grow and stay healthy. Female and male bodies look different from each other. This is partly because male and female hormones are different. The main female hormone is called **estrogen**. The main male hormone is called **testosterone.**

The Reproductive System

1. Fill in each blank with a term from the list. There will be one term left over.

| umbilical cord | sperm | fertilized | small | human | fetus |
| nine months | ninety days | navel | egg | belly button | |

Our reproductive system creates new _____ life. The female body
a

produces an _____ cell. It is _____ by the male's _____
b   **c**   **d**

cell. These two cells join together but are still very _____ . They grow to
e

become a newborn baby in _____ . The growing baby lives inside its
f

mother. Inside her body, it is called a _____ . The fetus gets its food through
g

a tube in its stomach. This tube is called the _____ . It is attached to the
h

inside of the mother, too. When we are born, the umbilical cord is cut off. This leaves

a spot on our stomach called our _____ or _____ .
i   **j**

2. Circle T if the statement is TRUE or F if it is FALSE.

T F **a)** Egg and sperm cells are the size of a pea.

T F **b)** Human babies are made through sexual reproduction.

T F **c)** Hormones are special chemicals that help us grow.

T F **d)** A fetus is another name for a developing baby.

T F **e)** A fetus does not need to eat until it is born.

T F **f)** The male reproductive hormone is called testosterone.

T F **g)** The female reproductive hormone is called estrogen.

After You Read

The Reproductive System

3. **How do the egg and sperm cells become a baby?**

_____

_____

_____

4. **How does the fetus breathe and get its food before it is born?**

_____

_____

_____

Extension & Application

5. We have read that **hormones** help us grow and stay healthy. We have also learned about two different kinds of hormones – estrogen and testosterone. Do some research to learn more about **what our hormones do**. Look for your information in books from the library or in an encyclopedia. List the names of the hormones that you read about. Record your information in a t-chart like this:

Hormone Name	What It Does

6. Write a **creative story** about seeing the world as a baby. How would you find out about the world? Use what you know about the senses to write this story.

Take Your Own Pulse

In this activity you will measure your pulse (heart rate). Your pulse is the number of times your heart beats in a period of time.

First, answer this question in your notebook:

When is your pulse faster – at rest or after exercise? Explain your reasoning.

FOR THIS ACTIVITY, you will need a stopwatch or timer.

STEPS:

1. (Sit down for this step.) Find your pulse by touching your finger to your wrist or the side of your neck. Ask your teacher for help if you need it.

2. Measure your pulse. Count the number of beats for 30 seconds.

    a) What is the number? _____

    b) Multiply this number by 2: _____ beats per minute. This is your pulse **at rest.**

3. Now, do one of the following: **jog on the spot** or **do jumping jacks** for 5 minutes.

4. Measure your pulse again for 30 seconds.

    a) What is the number? _____

    b) Multiply this number by 2: _____ beats per minute. This is your pulse **after 5 minutes of exercise.**

5. Next, sit down for 2 minutes. Then, measure your pulse again for 30 seconds.

    a) What is the number? _____

    b) Multiply this number by 2: _____ beats per minute. This is your pulse **after 2 minutes of rest.**

6. Sit down for 3 more minutes. Then, measure your pulse again for 30 seconds.

    a) What is the number? _____

    b) Multiply this number by 2: _____ beats per minute. This is your pulse **after 5 minutes of rest.**

    Record your results on a bar graph.

Look back at your answer to the first question. Was your guess correct? What conclusion can you make about pulse and exercise? Why do you think this is so?

Build a Kidney!

We have learned that the kidneys filter blood. This is how they help clean wastes from the body. For this activity, you will build your own "kidney". It is really some pop bottles, but it will give you an idea of how a filter, like our kidneys, works.

FOR THIS ACTIVITY, you will need:
- **3 large plastic pop bottles, cut in half (you will use the top half of each)**
- **a large bowl or bucket • pebbles • sand • paper towel • masking tape**
- **small jug of dirty water**

Ask your teacher to help you find these things.

STEPS:

1. Fill the first bottle with pebbles. Fill the second bottle with wet sand. Fill the third bottle with paper towel.

2. Stack the containers like they are in the picture and tape them together.

   They are now in a column.

3. Hold the column of containers over the bucket. Pour the dirty water over the pebbles in the top bottle. The water should filter all the way down into the bucket.

QUESTIONS:

Answer these questions in your notebook.

1. What does the water in the bucket look like? How is it different from the dirty water you started with?

2. What did the pebbles, sand and paper towel do?

3. What conclusions can you make about how kidneys work?

Diagram labels: dirty water in · rocks · sand · paper towel · pure water out

Organ System Poster

We have learned about FOUR important systems in the human body:

- **the circulatory system**
- **the digestive system**
- **the excretory system**
- **the reproductive system**

For this activity, you will **create a poster** with important facts about one of these systems.

YOUR POSTER SHOULD INCLUDE:

- a catchy title to get the reader's attention

- an illustration that shows what the system looks like (be sure to label all the parts!)

- the main parts that make up the system

- the main job or jobs of the system

- any other important information you can think of that can be presented in an interesting way

Begin by collecting important **facts** about your system. You may use the reading passages, the Internet, or other resource materials to find your information. Make your poster on a piece of Bristol board. Make it colorful, neat and organized.

When you have finished, **share** your poster with the class.

Pin the Organ on the Body

Here is an outline of the human body. To the left are pictures of important **ORGANS** in the body. Your task is to **CUT OUT** each organ and to **PASTE** it on the body where it belongs. You may use information from the reading passages, the Internet, or other resource materials to find the answers. (Hint: some of the organs may overlap!)

a) liver

b) intestines

c) esophagus

d) bladder

e) heart

f) lungs

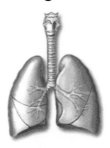

g) stomach

h) kidneys

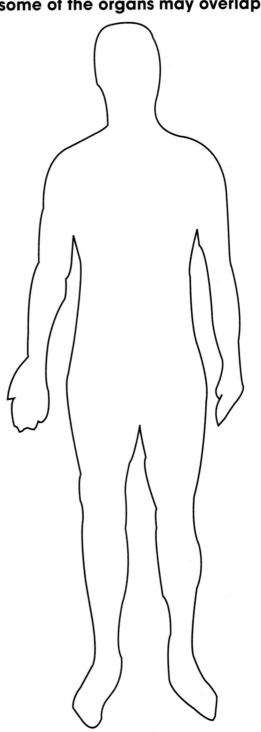

Crossword Puzzle!

Across

2. Blood _____ through the blood vessels

4. Another word for large intestine

6. Something with oxygen taken away

8. Food goes down this tube to the stomach

10. _____ blood cells help with our immunity

11. Organs that help us breathe and get rid of waste

12. This is removed from waste in the large intestine

15. This organ is coiled like a snake

17. The main job of the liver is to ____ blood

18. Small holes in our skin that sweat comes out of

DOWN

1. Big artery attached to the heart

2. Enzymes are _____ in the small intestine

3. A chemical in our stomach that helps us digest

5. A main organ of the circulatory system

7. These take blood away from the heart

9. Estrogen is a _____ made in the female body

13. Blood clots when it _____

14. Holds urnine

16. Largest organ of the body

Word List

deoxygenated,	hormone,
circulates,	thickens,
water,	acid,
pores,	chemicals,
esophagus,	veins,
lungs,	bladder,
colon,	skin,
white,	aorta,
small intestine,	heart,
filter	

Word Search

Find all of the words in the Word Search. Words are written horizontally, vertically, or diagonally, and some are even written backwards.

circulate
artery
vein
urine
capillary
pressure
excretory system
oxygen
bladder
digestion

fetus
involuntary
heart
chamber
nutrients
abdomen
plasma
platelet
clot

immunity
carbon dioxide
saliva
stomach
esophagus
kidneys
intestine
colon
enzymes

A	W	S	T	Z	C	S	D	F	T	G	B	N	H	Y	U	M	K
M	K	L	Y	F	H	E	A	R	T	R	F	V	S	D	R	E	X
C	S	T	O	M	A	C	H	H	C	I	R	C	U	L	A	T	E
Z	A	Q	X	S	M	R	F	V	B	N	X	D	E	R	H	S	F
C	D	E	Y	F	B	T	G	B	N	T	U	J	M	K	N	Y	I
N	P	T	G	U	E	J	K	L	N	E	W	F	E	T	U	S	N
W	L	R	E	Y	R	K	Y	U	G	S	O	V	X	D	E	Y	V
Q	A	S	N	U	T	R	I	E	N	T	S	E	J	F	T	R	O
M	S	H	Y	T	G	B	F	D	E	I	T	I	U	R	P	O	L
V	M	D	F	H	U	H	V	T	N	N	C	N	S	E	G	T	U
X	A	E	D	C	V	F	T	Y	G	E	C	V	N	D	G	E	N
U	H	V	B	E	N	Z	Y	M	E	S	Y	R	F	D	V	R	T
R	P	L	A	T	E	L	E	T	H	A	G	S	X	A	H	C	A
Y	R	E	T	R	A	R	V	G	Y	L	R	F	I	L	A	X	R
R	E	D	V	B	U	Y	U	O	L	I	F	D	M	B	B	E	Y
A	V	F	R	S	N	H	Y	R	F	V	D	G	M	S	D	K	J
L	W	C	S	V	F	R	T	G	H	A	U	J	U	R	O	F	K
L	S	E	S	O	P	H	A	G	U	S	N	Y	N	B	M	D	F
I	R	A	C	A	R	B	O	N	D	I	O	X	I	D	E	U	Y
P	U	H	N	T	V	R	F	T	G	B	L	N	T	H	N	O	L
A	Q	S	O	X	E	D	C	R	F	V	O	T	Y	G	B	N	H
C	M	L	J	U	U	R	I	N	E	K	C	I	L	O	P	S	F
X	C	V	D	E	R	F	D	G	N	O	I	T	S	E	G	I	D

Comprehension Quiz

Part A

34

Circle **T** if the statement is **TRUE** or **F** if it is **FALSE**.

8

T F   **1)** Blood circulates through blood vessels and goes to all parts of the body.

T F   **2)** The largest artery is the aorta, located in the lungs.

T F   **3)** The heart is a pump made of voluntary muscle tissue.

T F   **4)** Our stomach is not very strong. This is why we can get sick to our stomach.

T F   **5)** Acid in the stomach breaks down our food.

T F   **6)** Materials left over in the large intestine are indigestible.

T F   **7)** Sugar is a waste material found in sweat.

T F   **8)** Testosterone is an enzyme made in the male body.

Part B

Label the parts of the digestive system. Use the words in the list.

stomach   **esophagus**   **large intestine**   **small intestine**

8

1

2

4 _____ 3

SUBTOTAL: /16

Human Body CC4519

After You Read

Comprehension Quiz

Part C

1. Name the **three** kinds of **blood vessels**. Describe what each blood vessel does.

⑥

_____

_____

_____

2. Name **one** part of the **blood**. Describe its main job. Tell one other important thing about it.

③

_____

_____

_____

3. What happens to food when it enters the digestive system? Use the terms **teeth, esophagus, stomach** and **large intestine** in your answer.

④

_____

_____

_____

4. Name **one** organ of the **excretory system**. Describe what it does.

②

_____

_____

_____

5. How does a **fetus** get its food and oxygen? Why does it need these two things?

③

_____

_____

_____

SUBTOTAL: /18

2. Thin and long, round, rectangular

3. Like a factory, cells have many parts that work together to get important work done

4.
a) environment
b) hereditary
c) elements
d) DNA
e) protein

5.
a) Cell membrane b) Nucleus
c) Ribosome d) Lysosome
e) Mitochondria

6. Answers will vary

⑯

1.
cell membrane – **A, G**
lysosomes – **C, D/E/F**
cytoplasm – **B, J, L**
nucleus – **D/E/F, I, K**
mitochondria – **D/E/F, H**

⑮

1.
a) cytoplasm
b) nucleus
c) DNA
d) cell membrane
e) lysosomes
f) mitochondria

2.
a) cell
b) cytoplasm
c) cell membrane
d) nucleus

⑬

Cell membrane – lets some things in, keeps other things out

⑭

3. All living things are made of cells

4. It can do everything an organism needs to do to stay alive and healthy (eat, move and breathe)

5.
a)
b)
c)
d)
e)
f)

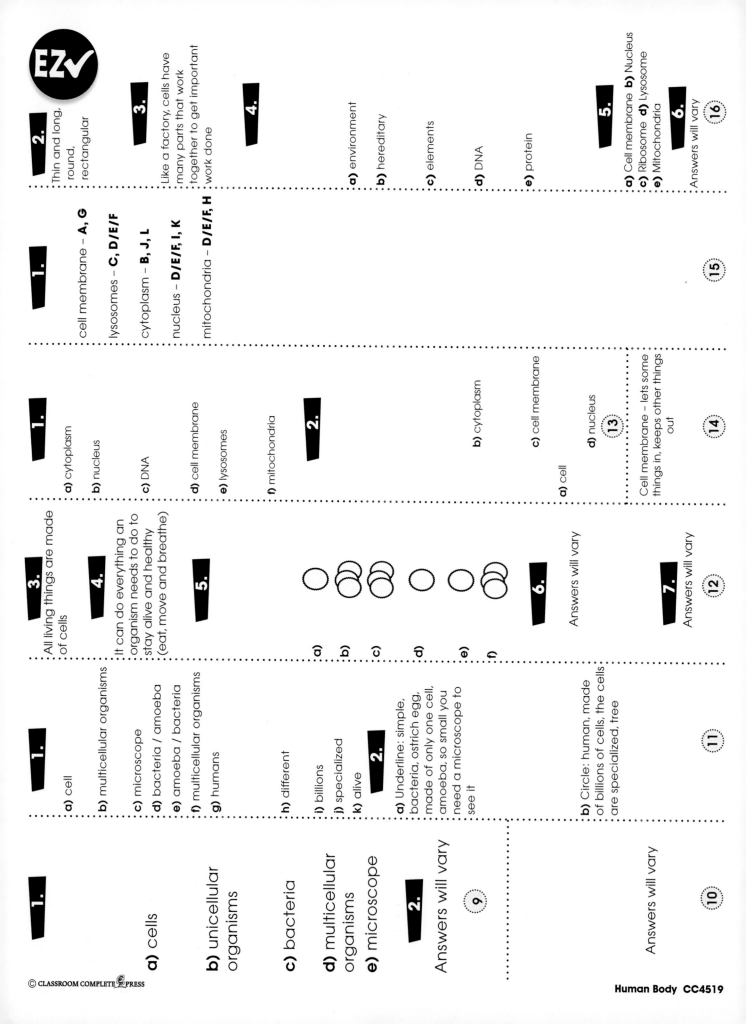

6. Answers will vary

⑫

7. Answers will vary

1.
a) cell
b) multicellular organisms
c) microscope
d) bacteria / amoeba
e) amoeba / bacteria
f) multicellular organisms
g) humans
h) different
i) billions
j) specialized
k) alive

2.
a) Underline: simple, bacteria, ostrich egg, made of only one cell, amoeba, so small you need a microscope to see it
b) Circle: human, made of billions of cells, the cells are specialized, tree

⑪

1.
a) cells
b) unicellular organisms
c) bacteria
d) multicellular organisms
e) microscope

2. Answers will vary

⑨

Answers will vary

⑩

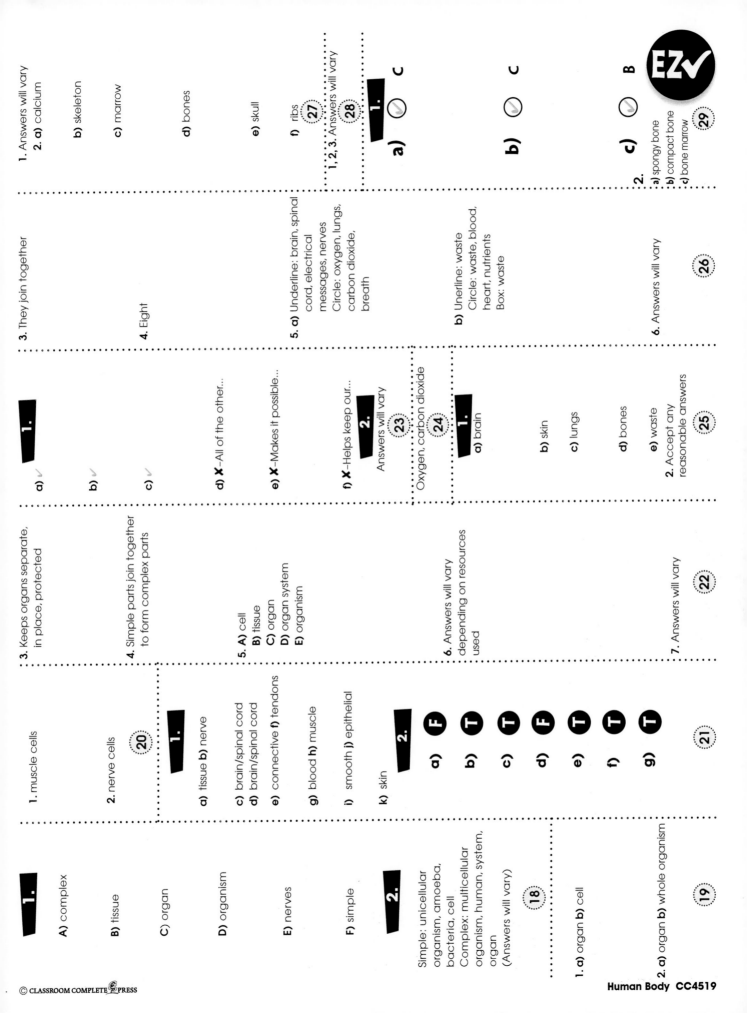

1. Answers will vary
2. a) calcium
   b) skeleton
   c) marrow
   d) bones
   e) skull
   f) ribs (27)

1, 2, 3. Answers will vary (28)

1. a) C
   b) C
   c) B (29)
2. a) spongy bone
   b) compact bone
   c) bone marrow

EZ✔

---

3. They join together

4. Eight

5. a) Underline: brain, spinal cord, electrical messages, nerves
   Circle: oxygen, lungs, carbon dioxide, breath

   b) Underline: waste
   Circle: waste, blood, heart, nutrients
   Box: waste

6. Answers will vary (26)

1. a) ✓
   b) ✓
   c) ✓
   d) X –All of the other...
   e) X –Makes it possible...
   f) X –Helps keep our...
2. Answers will vary (23)

Oxygen, carbon dioxide (24)

1. a) brain
   b) skin
   c) lungs
   d) bones
   e) waste
2. Accept any reasonable answers (25)

---

3. Keeps organs separate, in place, protected

4. Simple parts join together to form complex parts

5. A) cell
   B) tissue
   C) organ
   D) organ system
   E) organism

6. Answers will vary depending on resources used

7. Answers will vary (22)

1. muscle cells
2. nerve cells (20)

1. a) tissue b) nerve
   c) brain/spinal cord
   d) brain/spinal cord
   e) connective f) tendons
   g) blood h) muscle
   i) smooth j) epithelial
   k) skin

2. a) F
   b) T
   c) T
   d) F
   e) T
   f) T
   g) T (21)

---

1. A) complex
   B) tissue
   C) organ
   D) organism
   E) nerves
   F) simple

2. Simple: unicellular organism, amoeba, bacteria, cell
   Complex: multicellular organism, human, system, organ
   (Answers will vary) (18)

1. a) organ b) cell
2. a) organ b) whole organism (19)

3. Outside: compact bone
Middle: spongy bone
Inside: bone marrow

4. Accept any reasonable
answer

5. Heart, lungs, liver; Possible
fourth organ–stomach
(Answers will vary)

6. Answers will vary

7. Answers will vary

1.
a) joints
b) hinge
c) socket
d) rotation
e) cartilage
f) grinding

2.
Answers will vary

Answers will vary

(30)

(31)

(32)

1.
a) F
b) T
c) F
d) F
e) T
f) T
g) T
h) F
i) T
j) T

2.
a) sliding joint
b) ball and socket joint
c) hinge joint

(33)

3. They allow bones to move

4. It would be painful, difficult.
Answers will vary

5. Answers will vary

6. Answers will vary

7. Answers will vary

8. Answers will vary

9. Answers will vary

(34)

1.
A) fiber
B) bundled
C) striated
D) cardiac
E) involuntary

2.
muscles give our body
shape, muscles help us
move

3.
Answers will vary

(35)

Answers will vary

(36)

1.
a) F
b) T
c) F
d) F
e) T
f) F
g) T
h) F
i) T

2.
a) two
b) shape
c) move
d) run/breathe
e) run/breathe
f) cells
g) elastic
h) fibers
i) bundled

(37)

3. They give our body
shape and help
us move

4. Stripes

5. Answers will vary

6. Answers will vary

7. Answers will vary

8. Answers will vary

(38)

1. Answers will vary

2. Voluntary: biting an apple,
kicking a ball, walking to the
bus

Involuntary: your heart
beating, blood moving in
your veins, getting goose
bumps at a scary movie,
digesting an apple

(39)

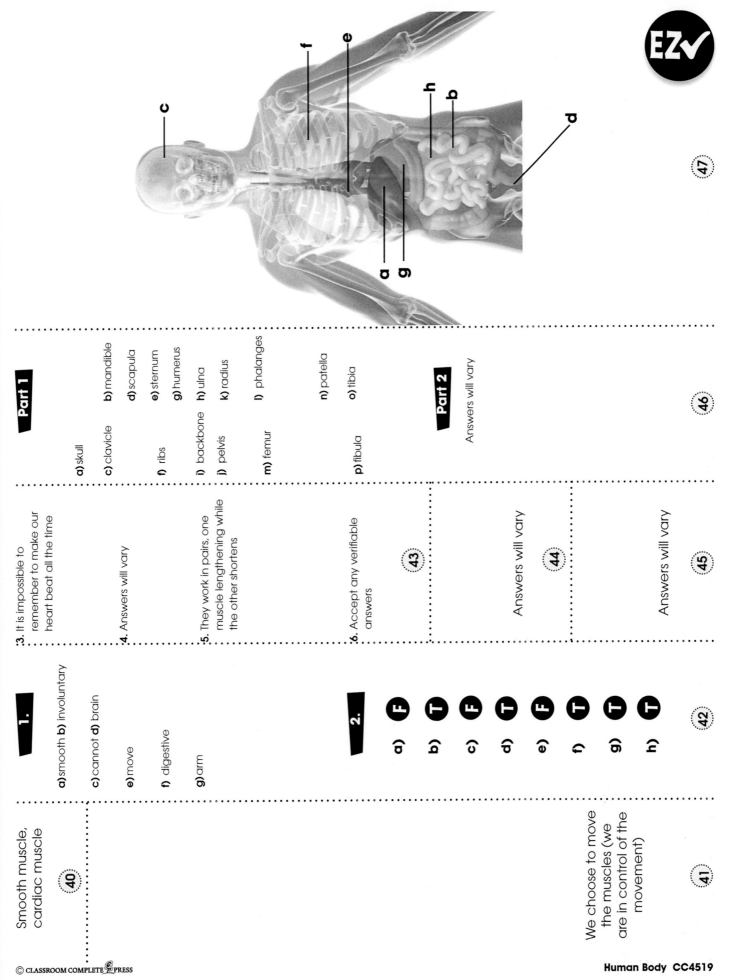

Diagram labels: c, f, e, h, b, a, g, d

Part 1

a) skull
b) mandible
c) clavicle
d) scapula
e) sternum
f) ribs
g) humerus
h) ulna
i) backbone
j) pelvis
k) radius
l) phalanges
m) femur
n) patella
o) tibia
p) fibula

Part 2

Answers will vary

(47)   (46)

3. It is impossible to remember to make our heart beat all the time

4. Answers will vary

5. They work in pairs, one muscle lengthening while the other shortens

6. Accept any verifiable answers   (43)

Answers will vary   (44)

Answers will vary   (45)

1.

a) smooth b) involuntary
c) cannot d) brain
e) move
f) digestive
g) arm

2.

a) F
b) T
c) F
d) T
e) F
f) T
g) T
h) T

(42)

Smooth muscle, cardiac muscle   (40)

We choose to move the muscles (we are in control of the movement)   (41)

Word Search Answers

Across:

1. contracting

3. specialized

4. cells

6. multicellular

8. hinge

10. skeletal

11. tissues

13. DNA

14. eight

Down:

1. cytoplasm
2. cartilage
4. cardiac
5. signals
7. involuntary
9. pairs
12. energy

Part A

1) **T**
2) **T**
3) **F**
4) **F**
5) **F**
6) **T**
7) **F**
8) **F**

Part B

1. spongy bone
2. compact bone
3. bone marrow

Part C

1. Cells that do specific jobs and not other jobs; multicellular; possible example: human

2. Answers will vary

3. Answers will vary

4. **Voluntary** – we can control it
Involuntary – we cannot control it. Examples will vary

5. **Possible answer:** We decide to move. The brain sends an electrical signal to muscles. The muscle pair moves (one contracts (shortens) and the other lengthens). Bone moves because muscles are working.

1.
a) brain
b) data
c) messages
d) spinal cord
e) nerves

2.
a) brain
b) spinal cord
c) nerves

(52)

Answers will vary

(53)

1.
a) brain b) control
c) protected d) skull
e) three
f) one hundred
g) cerebrum
h) cerebellum
i) brain stem

2.
Underline:
all *except* control center

3.
a) cerebrum
b) brain stem
c) cerebellum

(54)

4.
Answers will vary

5.
brain, spinal cord, nerves

6.
a) cerebrum
b) cerebellum
c) cerebellum
d) cerebrum
e) cerebrum
f) brain stem

7.
Answers will vary depending on resources used

(55)

1.
a) vertebra
b) neuron
c) spinal cord
d) brain
e) tissue

2.
a) protected

b) bundle

c) messages

(57)

For protection

(58)

1.
a) (circle) B

b) (circle) B

2.
a) neurons
b) messages
c) brain / spinal cord
d) spinal cord / brain
e) spinal cord
f) two
g) motor
h) sensory
i) hear j) smell

(59)

3.
Because it allows for communication between our brain and our body

4.
Answers will vary – activities must involve muscles

5.
a) vertebrae
b) nerve
c) protect
d) senses
e) neuron

6.
Answers will vary depending on resources used

7.
Answers will vary depending on resources used

EZ✔

(60)

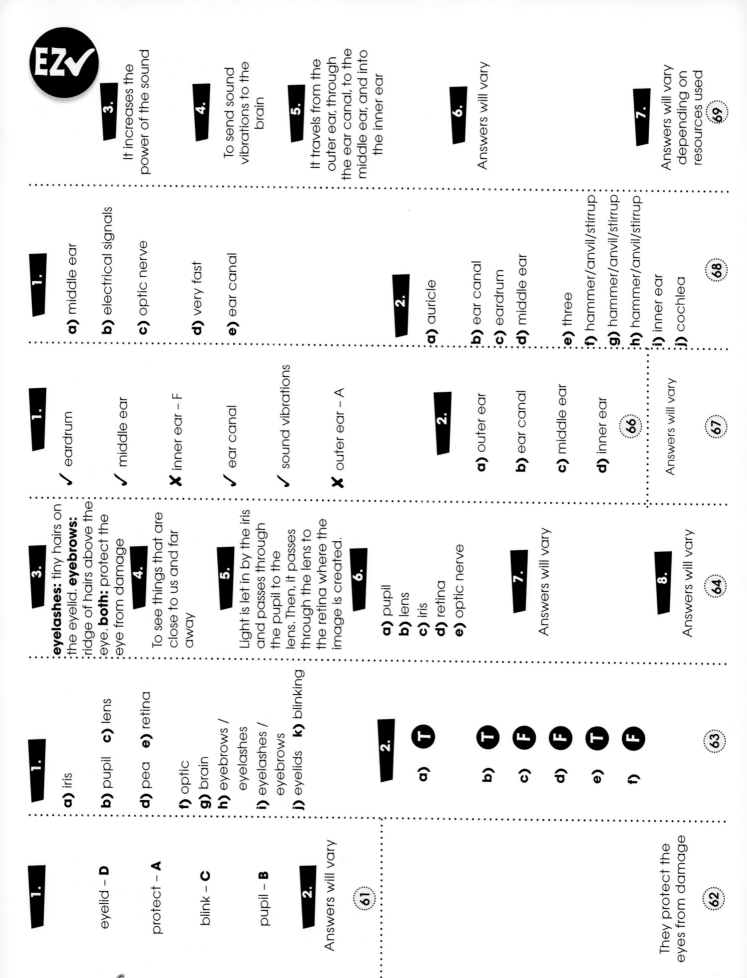

3. It increases the power of the sound

4. To send sound vibrations to the brain

5. It travels from the outer ear, through the ear canal, to the middle ear, and into the inner ear

6. Answers will vary

7. Answers will vary depending on resources used

(59)

1.
a) middle ear
b) electrical signals
c) optic nerve
d) very fast
e) ear canal

2.
a) auricle
b) ear canal
c) eardrum
d) middle ear
e) three
f) hammer/anvil/stirrup
g) hammer/anvil/stirrup
h) hammer/anvil/stirrup
i) inner ear
j) cochlea

(68)

1.
✓ eardrum
✓ middle ear
✗ inner ear – F
✓ ear canal
✓ sound vibrations
✗ outer ear – A

2.
a) outer ear
b) ear canal
c) middle ear
d) inner ear

(66)

Answers will vary

(67)

---

3. **eyelashes:** tiny hairs on the eyelid. **eyebrows:** ridge of hairs above the eye. **both:** protect the eye from damage

4. To see things that are close to us and far away

5. Light is let in by the iris and passes through the pupil to the lens. Then, it passes through the lens to the retina where the image is created.

6.
a) pupil
b) lens
c) iris
d) retina
e) optic nerve

7. Answers will vary

8. Answers will vary

(64)

1.
a) iris   c) lens
b) pupil   e) retina
d) pea
f) optic
g) brain
h) eyebrows / eyelashes
i) eyelashes / eyebrows
j) eyelids   k) blinking

2.
a) T
b) T
c) F
d) F
e) T
f) F

(63)

1.
eyelid – **D**
protect – **A**
blink – **C**
pupil – **B**

2. Answers will vary

(61)

They protect the eyes from damage

(62)

Across:

- **2.** train station
- **6.** motor
- **8.** iris
- **9.** lens
- **10.** brain stem
- **11.** canal
- **12.** survival
- **14.** vertebra
- **15.** connected
- **18.** optic
- **19.** cells
- **20.** bundle

Down:

- **1.** posture
- **3.** retina
- **4.** nasal cavity
- **5.** outer
- **7.** messages
- **13.** protected
- **16.** oxygen
- **17.** computer

Word Search Answers

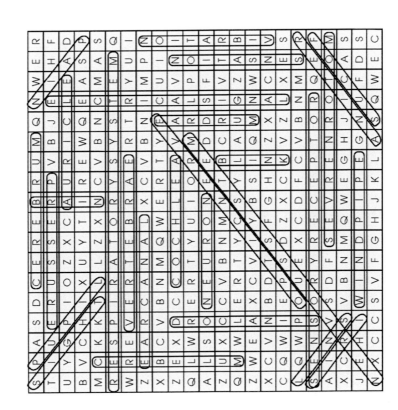

Part A

1) **T**
2) **T**
3) **F**
4) **T**
5) **F**
6) **F**
7) **F**
8) **T**

Part B

a) brain

b) spinal cord

c) nerves

Part C

1. 3 parts: brain, spinal cord, nerves. Accept any reasonable answers.

2. One of: cerebrum, cerebellum, brain stem. Accept any reasonable answers.

3. One of: motor nerves, sensory nerves. Accept any reasonable answers.

4. We breathe in (inhale) oxygen from the air into our lungs where it is carried to the bloodstream. Carbon dioxide is carried from the bloodstream to our lungs. Then we exhale it from our body (breathe it out). Answers will vary.

5. When something harmful touches our skin, we feel pain. Sensory receptors that feel pain are turned on and send a message to the brain. Feeling pain is important for our survival because it helps protect us from danger. Answers will vary.

91 92 93 94

3.

Answers will vary

4.

bright – oxygenated, going to cells
dark – deoxygenated, going to lungs

5.

a) pump

b) chamber

c) oxygenated

d) involuntary

e) deoxygenated

6.

Left side should be colored red.
Right side should be colored blue.

7.

Answers will vary

1.

a) Ⓐ B

b) Ⓐ C

2.

a) two

b) cells
c) oxygenated
d) bright e) lungs

f) dark

g) deoxygenated
h) poor

1.

a) oxygenated

b) chamber

c) involuntary

d) cardiac

e) pump

2.

a) beats

b) fist

c) cardiac

d) chambers

e) involuntary

(99)

To pump blood

3.

Due to pressure from the pumping of blood by the heart

4.

So they can drop off oxygen and pick up waste

5.

a) capillaries

b) arteries

c) veins

d) arteries

e) capillaries

f) veins

6.

Answers will vary

1.

a) circulatory b) food

c) waste d) circulates

e) heart f) tubes

g) blood

h) arteries i) veins
j) capillaries

2.

a) Underline: circulates, cleans out waste, brings oxygen to the cells, pumps blood, helps keep us healthy

b) Circle: arteries, blood, heart, vessels, aorta, capillaries, veins

1.

a) capillary

b) pressure

c) veins

d) circulates

e) aorta

2.

Answers will vary

(95)

Heart, blood, blood vessels

3. involuntary

4. Answers will vary

5. Answers will vary

6. Answers will vary

(112)

1. Answers will vary

2.
a) enzymes

b) diameter

c) indigestible

d) coiled

(113)

Answers will vary

(114)

1.
a) saliva

b) muscle

c) esophagus

d) four

e) fist

f) churns

2.
a) four

b) intestine **c)** mouth

d) teeth **e)** tongue

f) saliva

g) swallow **h)** starch

i) sugar

j) many

(111)

---

1. Answers will vary

2. Answers will vary

(109)

Teeth – bite, grind and rip food,

Tongue – moves food around,

Saliva – breaks starchy food into simple sugars

(110)

---

3. Answers will vary

4. In bone marrow

5. We would have less immunity, and would be more likely to get sick

6.
Scab – **B**

Plasma – **A**

Red blood cells – **C**

7. Answers will vary

8. Answers will vary depending on resources used

(108)

1.
a) four **b)** plasma

c) water **d)** red

e) iron

f) white
g) protecting
h) immunity
i) platelets
j) clots
k) scab

2.
a) F
b) T
c) T
d) F
e) T
f) T
g) F

(107)

1.
plasma – **C**

platelet – **A**

clot – **E**

volume – **D**

immunity – **B**

2. Answers will vary

(105)

Red blood cells, white blood cells, plasma, platelets

(106)

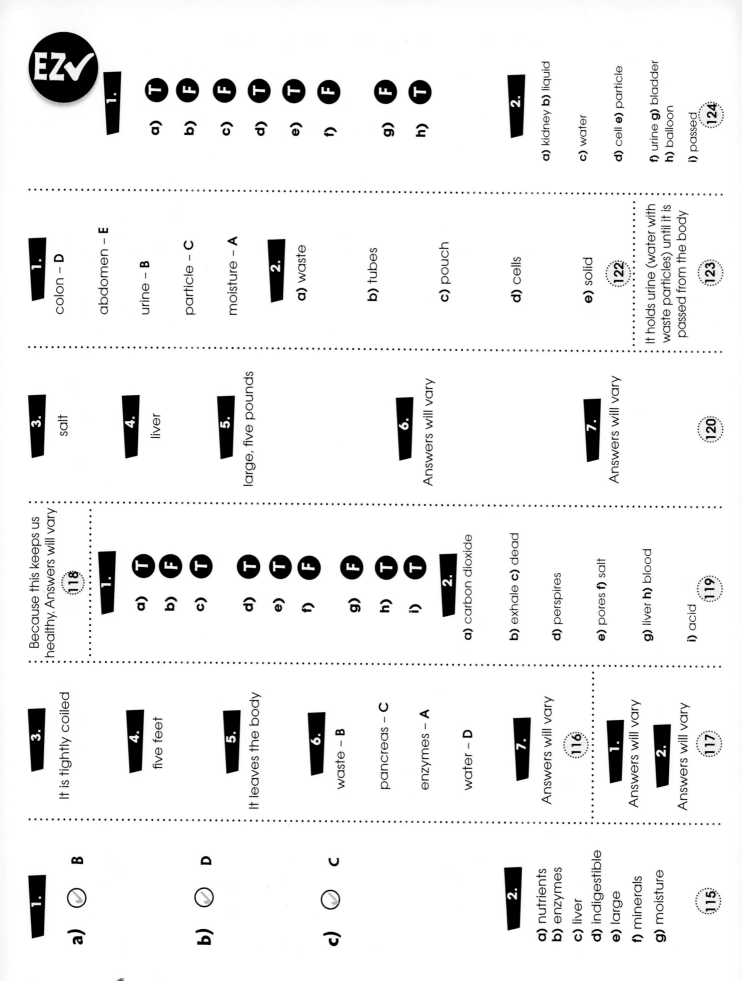

EZ✓

1.
a) T
b) F
c) F
d) T
e) T
f) F
g) F
h) T

2.
a) kidney b) liquid
c) water
d) cell e) particle
f) urine g) bladder
h) balloon
i) passed
124

1.
colon – D
abdomen – E
urine – B
particle – C
moisture – A

2.
a) waste
b) tubes
c) pouch
d) cells
e) solid
122

It holds urine (water with waste particles) until it is passed from the body
123

---

3. salt
4. liver
5. large, five pounds
6. Answers will vary
7. Answers will vary
120

Because this keeps us healthy. Answers will vary
118

1.
a) T
b) F
c) T
d) T
e) T
f) F
g) F
h) T
i) T

2.
a) carbon dioxide
b) exhale c) dead
d) perspires
e) pores f) salt
g) liver h) blood
i) acid
119

---

3. It is tightly coiled
4. five feet
5. It leaves the body
6. waste – B
pancreas – C
enzymes – A
water – D
7. Answers will vary
116

1. Answers will vary
2. Answers will vary
117

1.
a) B
b) D
c) C

2.
a) nutrients
b) enzymes
c) liver
d) indigestible
e) large
f) minerals
g) moisture
115

© CLASSROOM COMPLETE PRESS

Human Body CC4519

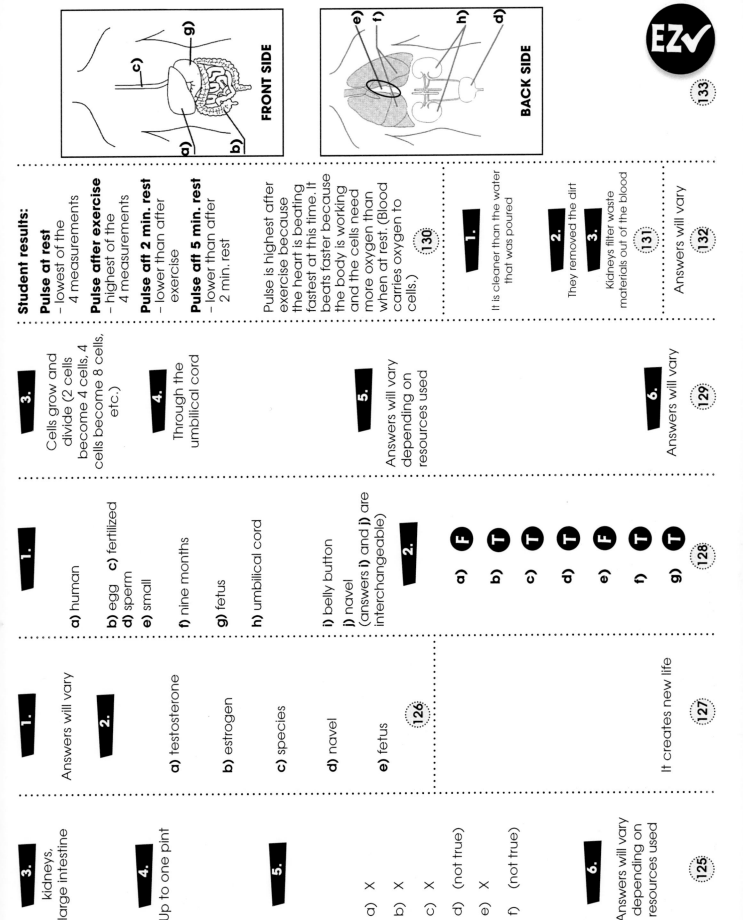

FRONT SIDE

BACK SIDE

133

Student results:

Pulse at rest
- lowest of the 4 measurements

Pulse after exercise
- highest of the 4 measurements

Pulse aft 2 min. rest
- lower than after exercise

Pulse aft 5 min. rest
- lower than after 2 min. rest

Pulse is highest after exercise because the heart is beating fastest at this time. It beats faster because the body is working and the cells need more oxygen than when at rest. (Blood carries oxygen to cells.)

130

1. It is cleaner than the water that was poured

2. They removed the dirt

3. Kidneys filter waste materials out of the blood

131

Answers will vary

132

---

3. Cells grow and divide (2 cells become 4 cells, 4 cells become 8 cells, etc.)

4. Through the umbilical cord

5. Answers will vary depending on resources used

6. Answers will vary

129

1.
a) human
b) egg c) fertilized
d) sperm
e) small
f) nine months
g) fetus
h) umbilical cord
i) belly button
j) navel
(answers **i)** and **j)** are interchangeable)

2.
a) **F**
b) **T**
c) **T**
d) **T**
e) **F**
f) **T**
g) **T**

128

---

1. Answers will vary

2.
a) testosterone
b) estrogen
c) species
d) navel
e) fetus

126

It creates new life

127

---

3. kidneys, large intestine

4. Up to one pint

5.
a) X
b) X
c) X
d) (not true)
e) X
f) (not true)

6. Answers will vary depending on resources used

125

Across:

2. circulates

4. colon

6. deoxygenated

8. esophagus

10. white

11. lungs

12. water

15. small intestine

17. filter

18. pores

Down:

1. aorta

2. chemicals

3. acid

5. heart

7. veins

9. hormone

13. thickens

14. bladder

16. skin

Word Search Answers

Part A

1) T

2) F

3) F

4) F

5) T

6) T

7) F

8) F

Part B

1. esophagus

2. stomach

3. small intestine

4. large intestine

Part C

1. arteries – take blood away from the heart
veins – take blood to the heart
capillaries – carry blood all over the body.

2. One of: red blood cells, white blood cells, plasma, platelets. Answers will vary.

3. Teeth chew food into small bits. It goes down the esophagus to the stomach where it is broken down even more. Then it goes to the large intestine before it leaves the body. (Answers will vary)

4. Answers will vary

5. Through the umbilical cord that is attached to its belly and the inside of the mother's belly. To help it grow and develop.

Human Body CC4519